D0566958

WITHDRAWN
Wilmette Public Library

WILMETTE PUBLIC LIBRARY
1242 WILMETTE AVENUE
WILMETTE, IL 60091
847-256-5025

MASTER
YOUR BRAIN

TRAINING YOUR MIND
for SUCCESS *in* LIFE

PHILLIP ADCOCK

WILMETTE PUBLIC LIBRARY

STERLING
New York

STERLING
New York

An Imprint of Sterling Publishing
1166 Avenue of the Americas
New York, NY 10036

STERLING and the distinctive Sterling logo are registered trademarks of
Sterling Publishing Co., Inc.

© 2015 by Phillip Adcock

All rights reserved. No part of this publication may be reproduced,
stored in a retrieval system, or transmitted in any form or by any means
(including electronic, mechanical, photocopying, recording, or otherwise)
without prior written permission from the publisher.

ISBN 978-1-4549-1605-5

Distributed in Canada by Sterling Publishing
c/o Canadian Manda Group, 664 Annette Street
Toronto, Ontario, Canada M6S 2C8
Distributed in the United Kingdom by GMC Distribution Services
Castle Place, 166 High Street, Lewes, East Sussex, England BN7 1XU
Distributed in Australia by Capricorn Link (Australia) Pty. Ltd.
P.O. Box 704, Windsor, NSW 2756, Australia

For information about custom editions, special sales, and premium and corporate purchases,
please contact Sterling Special Sales at 800-805-5489 or specialsales@sterlingpublishing.com.

Manufactured in United States of America

2 4 6 8 10 9 7 5 3 1

www.sterlingpublishing.com

CONTENTS

158
AD

B·J 10/8/15

PART FOUR: CREATE A MINDSET FOR SUCCESS

Introduction

CONGRATULATIONS! You've just taken the most important step to achieving more of everything you've always wanted in life.

If you've ever tried and failed to quit smoking, doggedly followed a diet and ended up heavier, or repeatedly crashed and burned on the dating scene, you've probably found yourself in the self-help section of your local bookstore or library. There, you likely waded through scores of titles, each promising to be better than the last at helping you achieve all that your heart desires. Sadly, 99 percent of the time they don't succeed.

Let's begin by saying that this isn't just another self-help book. It's an "owner's manual" for your brain. In the coming chapters, you'll learn to make effective and lasting changes in your life by using more of your brain in ways that will seem astonishing and yet intuitive. If you didn't notice on your way past the table of contents, this book has fifty-two chapters. To get the most from *Master Your Brain*, read the book all the way through once. Then go back and read one chapter a week, focusing on completing the exercises and implementing the advice.

This book is about empowering you to reach whatever goals you set for yourself—wherever, whenever, and however you want. You'll find out how millions of people have improved their lives by simply learning from real-life examples and using the techniques and tools revealed in this book. When you've finished reading it, you'll be well on your way to creating a more extraordinary you.

But who am I to be making such bold claims?

As a student of human behavior and psychology for the past thirty years, I've developed the tools to deliver actionable behavior goals and psychological insights to guide some of the largest corporate entities in the world. Shopping Behavior Xplained Ltd., the organization that I founded, specializes in studying the actions and habits of consumers. Our goal is to help retailers communicate with customers more effectively. Our analysis and interpretation of human behavior have led to sales growth, improved service, increased customer satisfaction, and a wealth of other benefits. In short, I know humans. That's my bottom-line business. I know how we think and why we behave the way we do. Unlike pseudo-psychologists and so-called mind experts, I have tested my hypotheses rigorously over the past three decades. I use that knowledge to advise the movers and shakers of tomorrow through public speaking and university lectures designed to lead people toward more productive lifestyles.

During my professional development, I embraced and incorporated the teachings of some of the world's leading psychologists. This book resulted from reading and digesting hundreds of scholarly books and thousands of academic papers on human psychology and behavior. Whether you have a single issue you'd like to address or you want to develop an entirely new mental strategy for your life, this book contains proven strategies to help you achieve your goals easily and effectively.

Think about that for a moment.

Now—what are you waiting for?

PART ONE

SAY HELLO
TO YOUR
BRAIN

✴

The Human Supercomputer

SCIENTIST and award-winning author Ray Kurzweil once wrote: "The human brain is capable of making an astonishing 20 million-billion calculations per second." That's power.

Our brains consist of around 73 percent water. A newborn baby's brain weighs between 10 and 14 ounces. It will grow to between 46 and 49 ounces by adulthood. On average, our brains comprise only 2 percent of our total body weight but consume around 20 percent of all the energy we use, and between 15 and 20 percent of all the blood pumped from our hearts heads straight to our brains to supply them with the energy they need. A normal adult human brain consists of around 100 billion neurons, the basic information-processing structures that make up this amazing organ. They connect to one another via synapses, through which chemical information flows from one neuron to another. This information passes between neurons by way of tiny chemical reactions in each neuron that fire an electrical or chemical signal across the synapse to the next neuron. Trillions of these impulses take place every second.

Neural pathways within the brain develop before birth. As a growing child associates images with words, the messages that cross these pathways become routine. With increased use, the correlation between a particular sound, such as "mama," and the sight of a face becomes established. As these pathways develop, they collectively become the map of how a person thinks, reasons, and remembers. The more often a person uses the same neural pathway, the stronger and more embedded that behavior becomes. That's why, for example, it's so difficult to break habits such as nail biting. On the flip side, practice makes perfect because it strengthens neural pathways, and that's how we improve ourselves.

Our brains can carry out incredible numbers of calculations in a remarkably short period of time. A neuron can fire approximately once every five milliseconds; that's about two hundred times per second. The number of cells to which each neuron connects varies, but each neuron connects to at least one thousand other neurons. So every time a neuron fires, at least one thousand other neurons receive information from that neuron. Do the math, and the number of calculations per second is astonishing:

100 billion neurons × 200 signals per second × 1,000 connections per signal = 20 quadrillion calculations per second

A typical 2 GHz computer processor can handle a mere two billion calculations per second.

But, as we'll discover, although this remarkable organ between our ears can perform almost unfathomable numbers of calculations in a single second, it isn't as advanced as we want to believe. Try this exercise:

<hr>

Exercise
The Beach

Visualize waves washing onto a beach. Add the sound of the waves crashing into the sand as they ebb and flow. Now conjure the smell of the salt air.

When you try to focus on all three components at once—sight, sound, and smell—your brain can't do it . . . at least not without outside assistance.

<hr>

Even though we all possess our own version of the fastest computer on Earth, we still need to learn more about our brains if we want to maximize their power.

REMEMBER . . .

* Practice strengthens desired neural pathways.
* The human brain can perform twenty quadrillion calculations per second, but we can focus on only one or two things simultaneously.

{2}

Evolution Is Everything

ACCORDING to the *Oxford English Dictionary*, evolution is a process by which different kinds of living organism are believed to have developed from earlier forms during the history of the earth. The leading proponent of evolutionary theory was Charles Darwin, an English naturalist and the author of *On the Origin of Species* (1859). His theory of evolution holds that all life is related and descends from a common ancestor through undirected "descent with modification." That is, complex creatures evolved naturally from simpler ancestors over a long period of time. Natural selection, also known as survival of the fittest, drives evolution and explains the incredible diversity of the natural world, including why certain species develop a particular sense—such as sight, hearing, or smell—above others. In our case, evolution explains why humans walk upright on two legs and possess remarkably dexterous hands and fingers.

Darwin and many scientists since concur that animals evolved their senses to ensure survival. Dogs, for example, have a highly acute sense of smell. Birds of prey have extraordinary eyesight. Bats have particularly keen hearing. Humans have better eyesight than any of our other senses, but we haven't developed it significantly above the others. Our sense of smell is the least beneficial to our survival, so the part of the brain known as the amygdala—which plays a primary role in emotional reactions, including the fight-or-flight response—has sidelined it. But another section of our brains has developed more than in other animals: the neocortex, which largely governs reasoning and constraining unacceptable instincts and emotionally driven behaviors.

We have evolved differently from other animals because we can evaluate situations. We can choose between alternative outcomes much better than other members of the animal kingdom. But whether we actively choose our actions or we rationalize and justify our instincts remains open to debate. Plus, having the ability to choose doesn't mean we always choose wisely.

Any study of human evolution begins with the Catarrhini: flat-nosed primates who are probably our ancestors dating back some thirty million years ago. They lived predominately in trees. Around six million years ago, those ancestors descended from the trees onto the plains of Africa. At the time, they were still quadrupeds ("four-footed") and moved around like most animals, meaning they couldn't look down on their surroundings or see over the tall pampas grasses around them. In response, our predecessors evolved to stand on their hind legs so they could have a better view—much as meerkats do today. They could then spot potential meals, approaching threats, and possible mates. Standing on two legs became an advantage to the human species. Unlike other primates, we *Homo sapiens* walk upright. This posture makes demands on the structures of our inner ear that, together with the brain, help us maintain our balance. The stabilizing organs in our brains originally evolved to balance us on all fours, which helps explains why you'll often tip your head thirty degrees forward to regain your balance, particularly on an uneven surface.

Resisting the evolutionary (and smartphone-driven) urge to walk with your head down, staring at the ground, offers a great way for you to improve your state of mind in an instant. For example, imagine that you're entering a room full of people with your head down. What you're communicating to others as well as to yourself is subservience. By averting your gaze and looking below others' sightlines, you're deferring to their dominance. Your brain detects this physiological state and provides the mindset to match: You feel unsure, less confident, and even intimidated by the situation.

To avoid this debilitating state, concentrate your vision on a height about as high as people's foreheads or just above their heads even before you enter the room. You'll be amazed at how many people will glance at you before they look down, subconsciously submitting to you. How's that for a quick and easy confidence builder?

Now what about talking? Scientists are less sure about when our ancestors first developed the ability to speak, and there are numerous hypotheses about how, why, when, and where language might have emerged. Natalie Thais Uomini and psychologist Georg Friedrich Meyer recently estimated that our ancestors first began to speak around 1.75 million years ago. Whatever the actual timescale, speech has formed a part of human communication for only a fraction of our evolutionary development, so our brains aren't automatically wired to think in terms of words, sentences, and the like. When we see or hear a word, our brains need to convert that word into something more meaningful, such as a mental image. Hence, the famous saying "a picture's worth a thousand words." This neurological point also demonstrates how ineffective written communication can be. If you want to communicate meaningfully and with emotion (more on that later), get in front of the person with whom you want to communicate.

On the other hand, we humans have been processing visual images for around ten thousand times as long as we've been using words. That's why so many ad campaigns rely on graphics to convey their messages.

Now apply that advertising strategy to yourself. People can't resist looking at other people, so include a flattering picture of your face on your résumé and other important communications. In a stack of a hundred or more résumés, the one with a picture will stand out.

The changes underlying human evolution come predominately from fluctuations in population growth. But the diverse terrains, climates, and social structures that we've experienced since the end of the last ice age, around twelve thousand years ago, also have affected our genetic inheritance. According to Stephen Jay Gould—paleontologist, evolutionary biologist, scientific historian, and one of the most widely read science writers of our time—human civilization arose with essentially the same bodies and brains that we've been using for the last forty thousand years. Recent research by anthropologist John Hawks of the University of Wisconsin at Madison suggests, however, that people today are genetically more different from people living five thousand years ago than those people were from the Neanderthals, who vanished thirty thousand years earlier. In other words, the rate at which we are evolving is increasing.

Nevertheless, evolution is an extremely slow process. The human race is still acclimating to the twenty-first century. As much as we may hate to admit it, we rely on the instincts of our hunter-gatherer ancestors to get us through each day.

REMEMBER . . .

* A picture really is worth a thousand words; include a flattering photo on your résumé.
* When entering a room full of people, focus on or above their foreheads to demonstrate confidence.
* To be the most effective, speak face-to-face for important communications.

{3}

The Power of Instinct

THE first vertebrates (species with backbones) appeared on Earth some six hundred million years ago. Modern humans—*Homo sapiens* with a similar skeletal build, including braincase and jaw anatomy—have been around only for a couple of hundred thousand years. That means that 97 percent of our evolution occurred before we emerged. In evolutionary terms, we're late to the party.

In his entertaining book *Kluge: The Haphazard Evolution of the Human Mind,* research psychologist Gary Marcus notes that our ancestral brain system has been evolving for hundreds of millions of years, but cognitive reasoning, a more modern development, began taking shape as recently as 1.5 million years ago. So the part of our brain that allegedly makes us more intelligent than other animals has only been operating for a fraction of the time that our brains have been evolving. And as we'll see later, this "intelligent" part of our brains can't compete with the older, more established, more evolved parts of our animal brains. In other words, almost every meaningful decision we make depends on the thought process used by the rest of the animal kingdom. That animal thought process is what we call instinct.

The point of our instincts is to preserve life and health, but instincts adapt to changing environments slowly. For example, our brains have a survival mode that compels us to hoard nutrients in case of a food shortage. In the hunter-gatherer days, humans rarely knew when their next full meal would come along, so they gorged themselves whenever they could but went for long periods between meals. In the modern, developed world, food shortages are rare, but the instinctive act of gorging continues. We're hardwired to eat when we see food and to eat more than we need to survive in our new food-rich societies. Occasionally, we'll hear on the news of an impending food shortage of one commodity or another—pecans, olive oil, etc.—resulting from some industrial action or problems in the supply chain. As a result, most of us rush out and buy pecans and olive oil even if we weren't already craving them. Another example: Notice how much food we buy right before a storm, blizzard, or Christmas—often far more than our families could possibly eat. We know that supermarkets will close, which instinctively drives us to stock up for the (relatively short) food-supply interruption.

Why We Make Bad Decisions

Have you ever wondered why our wonderfully efficient brains make dumb mistakes? Why do we accidentally blurt out secret information or hurtful words? Why do we touch that hot dinner plate in the restaurant right after the server warned us that it was piping hot?

A multitude of different components in our bodies send pieces of what becomes a mental representation to the center stage in our heads, but subconscious mental processing accounts for 95 percent of our decision-making. We're unaware of nearly all the decisions our brains are making for us. John Haugeland, a professor of philosophy at the University of Chicago, explains the 95/5 split between unconscious and conscious activity in the brain.

> Compared to unconscious processing, conscious thinking is conspicuously slow and laborious—not a lot faster than talking, in fact. What's more, it is about as difficult to consciously entertain two distinct trains of thought at the same time as it is to engage in two distinct and different conversations at once; consciousness is in some sense a linear or serial process in contrast to the many simultaneous cognitions that are manifest in unconscious action.

In other words, we often arrive at decisions before our brains have concluded which decision is best.

In the twenty-first century, we live our lives substantially faster than our brains can handle. They can deal easily with "fight, flight, or fornicate" (more about this later), but as we continue to improve technology and technology continues to revolutionize our lives, we're leaving ourselves behind. Think about how many times you reply to a text message before giving sufficient thought to what the original message was intended to mean and how best to respond.

Another issue is even scarier: When our brains don't know how to respond to a stimulus, they just guess. Only occasionally and with conscious, cognitive input do we gather all available information so we can make informed decisions. But if the knowledge base within our brains has gaps, we're prone to blind prediction to plug in those gaps.

If your eyes are following a moving object traveling in a straight line, your brain is constantly plotting where the object is heading; it's literally thinking ahead. But if the object suddenly changes direction, the brain has to recalculate the trajectory, which takes considerable time and effort. At that critical moment, if the brain is concentrating on something else (lunch, for instance), it will often blindly follow its original prediction—at least for a moment or two. That's one of the common reasons that traffic accidents happen. It's also why referees and umpires make so many wrong calls in sports.

Dieting in particular goes against our survival instinct to eat more than we need when food is available, which is one of the main reasons that nearly all diets fail. Yet in the United States alone, consumers spent $61 billion on weight loss in 2010. Typical dieters make four attempts a year to lose weight, which means they fail at least three times a year. If they understood that their brains were entering survival mode every time they started a new diet, they'd know how to combat this irresistible urge to eat (which we'll examine in detail later).

The inescapable truth is that we humans, like much of the animal kingdom, are at the mercy of our own instincts. The brain is the most complex, remarkable organ in nature, and, although every species of reptile, amphibian, and mammal has one, the human brain is still the most developed. But what makes the human brain special?

REMEMBER . . .

* Subconscious mental processing accounts for 95 percent of our decision-making.
* When time or information is scarce, our brains often guess, leading to hasty, ill-informed decisions.
* Like all animals, we base most of our decisions on instincts designed to preserve our health and well-being.
* We are hardwired to eat as much as possible when food is available, which is why nearly every diet fails.

⟨4⟩

Battle of the Brains

ONE part of our brains is more advanced and developed than in any other animal: the neocortex and more specifically the prefrontal cortex (PFC). It's at the very front of your brain, right behind your forehead. The PFC analyzes our thoughts and regulates our behavior. It's the part of the brain that helps us decide what's morally and socially right or wrong. It aims to predict likely outcomes from a proposed behavior while allowing us to think ahead, plan for the future, and adjust to ever-changing situations. The PFC also regulates the influence of powerful instincts and emotions generated by older, more developed parts of the brain. So part of our brains is designed specifically to prevent us from behaving badly, making inappropriate decisions, and acting antisocially. That's good, then, right?

Not quite. Another part of the brain has been around for much longer: the insula. Scientists largely ignored this part of the brain for a long time, but recently it has taken its rightful place at the center of why we humans feel human. Whereas the PFC is all about reason, the insula processes and responds to emotions and instincts. It's believed to be the source of social emotions, such as empathy, guilt, and pride, and it also responds to physiological states like hunger and thirst. It urges you to eat when you're hungry and drink when thirsty.

According to Martin Paulus, a psychiatrist at the University of California at San Diego, the insula is the part of the brain that integrates mind and body. The insula receives information from the body and then instructs other parts of the brain—mostly those involved in decision-making such as the PFC—to respond accordingly. Think of the insula as the mission control center that constantly monitors how you feel and then generates urges and cravings.

The insula tells the PFC what to do, but, like a petulant teenager rebelling against his parents, the PFC doesn't like being told what to do.

According to Antonio Damasio, a neuroscientist at the University of Southern California, rational thinking can't be separated from feelings and emotions, and the insula, he says, plays a starring role.

So what are the ramifications of having two controlling parts of our brains that rarely agree when it comes to decision-making? First, the older and more established insula usually wins. We prefer immediate gratification. What the insula wants, it wants *now*. The PFC then shifts into overdrive, trying to negotiate a more considered and acceptable course of behavior.

Imagine you're in a store and you see a pair of must-have shoes. The insula, which operates in the present and rarely thinks about next week, tomorrow, or even later today, is responsible for the "must have" part of the urge. It urges you to pull out your wallet and throw caution to the wind. At the same time, the PFC, which does think ahead, is screaming, "No, no, no!" It rationalizes that you'll have to put the shoes on your credit card and pay an exorbitant interest fee, and you already have more shoes than you need. This heated brain argument is a battle between emotion and reason, and unfortunately emotions win most of the time.

Our insulae make many misplaced decisions for us partly because the PFC hasn't evolved for as long a time and because it manages functions more essential to our survival. The dominance of emotion can damage our chances for success, but all is not lost. The key to mastering the insula is twofold: First recognize that emotional urges are driving your decision-making processes. Then introduce strategies that level the playing field between the PFC and the insula.

When faced with an urge to buy those shoes, tell your insula that you'll think it over for twenty-four hours and you'll buy them then if they're still a must-have. This strategy helps in two ways: First, it provides the insula with an emotionally satisfying solution because you've agreed to buy the shoes. But second, it gives the PFC time to assemble a stronger case for the defense. If after a day the shoes still seem like a necessity, chances are good that they might be a worthwhile purchase. That way, your decision will have come from a combination of emotional and rational thought.

So the next time you find yourself facing an irresistible urge to buy or say or do something, give reason a chance to catch up with emotion and instinct. If that doesn't work, try playing the Why, Why, Why game: *Why* did that person offend me? *Why* do I feel the urge to respond? *Why* will I benefit from not reacting?

REMEMBER . . .

* Emotional impulses drive your decisions.
* Practice delayed gratification to level the playing field between the warring parts of your brain.

{5}

Fight, Flight, or Fornicate

RESISTING the impulse to buy shoes is one thing, but the insula can cause problems in other ways. For instance, you might see someone you'd like to get to know better but can't work up the courage to make the first move. Your insula is protecting you from the possibility of rejection or embarrassment. In that case, you need an approach strategy with which you can call on your PFC to help overcome those negative, insula-inspired fears.

Our brains haven't evolved in a way that benefits us very much in the twenty-first century. They evaluate three basic parameters whenever we encounter sensory stimuli. Whether we see, hear, smell, touch, or taste something, our initial mental response is to run it by the powerful filtering mechanism of the three Fs: fight, flight, or fornicate.

The hardwired fight-or-flight response protects us from bodily harm. The "fornicate" part of the equation came along a little more recently, but it's an equally vital response option. When our brains trigger one of the three Fs, our bodies tense, become more alert, and act more quickly. If a mugger is approaching you, you need to take action—and *fast*. Parts of the brain known collectively as the sympathetic nervous system send out impulses that in turn send more adrenaline rushing into the bloodstream, which causes physiological adjustments in your body. These changes include an increase both in your heart rate and blood pressure. At the same time, another part of your brain orders the release of numerous hormones that ready your body to deal with the perceived threat.

In addition to increased heart rate and blood pressure, the level of glucose in your bloodstream increases and is diverted toward your muscles, enabling them to take responsive action. Some nonessential systems, such as digestion, close down to allow more energy for fight-or-flight functions. Other changes include the tensing of muscles (which often leads to goose bumps), dilation of the pupils to take in as much light as possible, and inhibition of hearing and peripheral vision to allow your brain to focus on the impending threat.

The urge to fornicate is also an autonomic response and nearly impossible to prevent from happening. But that scenario usually happens with fewer extreme bodily changes and often only after the fight-or-flight response has occurred. When you experience the third F, you know you're going to fall in love or lust.

Rutgers University anthropologist Helen Fisher describes the human body as a finely tuned attraction-seeking machine that requires only a fraction of a second to determine the attractiveness of someone. When we spot a face or body that attracts us, our bodies produce a wide range of physical signs that communicate our attraction. Our brains release dopamine, which causes our pupils to dilate and our palms to sweat. Our hearts beat faster as our brains spur us on to make that first move. Even before you have time to process the situation, your body has already prepped you for action.

————

The brain has developed fantastic strategies to help us survive, but not all are suitable all the time. If you see a runaway truck heading toward you, you'd better act instantaneously and without running the information past your prefrontal cortex. But if you're not in immediate danger, you can improve your decision-making process by getting a second opinion from the PFC.

The fight, flight, or fornicate response looms behind pretty much every decision we make. For example, when a stranger approaches you in a bar, you brain automatically seeks cues to whether your best response is to fight, flee, or fornicate. The same happens when your boss calls out your name across the office. In a store, think about when somebody bumps into you or even stands a bit too close. That's your brain evaluating the situation.

The next time you sense your brain summoning the three Fs over something less critical than impending disaster, ask yourself what's causing the sweating palms and racing heartbeat. Is it really a three-F moment, or is your brain conflating mortal danger with potential embarrassment at speaking to your colleagues? As you grow increasingly comfortable with the three-F response, try to attune yourself to times when your brain misinterprets the "threat."

Exercise
Shopper's Paradise

If you're on edge, here's a way to calm your automatic reaction to approaching strangers, unexpected encounters in public, or that uncomfortable feeling when someone invades your space. Diffuse your emotional instincts by focusing on how you feel. Are you suddenly tense, excited, or nervous? The act of focusing on your emotional state will alter that mindset because you're allowing your PFC to put your feelings into context. A grocery store or mall, for example, is quite a safe location, and the three Fs seldom takes place in either.

Becoming aware of how you feel and then understanding the possible causes allow you to calm your physiological reactions, thus reducing the intensity of that emotional state.

REMEMBER . . .

* Your brain is wired to react to potential threats by releasing dopamine, which triggers the "fight, flight, or fornicate" response.
* If not faced with immediate danger, stop and allow your prefrontal cortex to reassess the situation.

{6}

Mirror Neurons

I N the 1990s, a team of neuroscientists at the University of Parma made a surprising discovery: Certain groups of neurons in the brains of macaque monkeys fired not only when a monkey performed an action—grabbing a banana, for example—but also when the monkey watched another monkey performing the same action. These neurons even fired when the monkey only heard another monkey performing the action in another room. The scientists realized that certain neurons in the monkeys' brains were "mirroring" the behavior of the other monkeys.

From an evolutionary perspective, mirror neurons are thought to have developed in order to help animals learn more quickly. For instance, the sooner a monkey learns to tell the difference between a banana and a venomous yellow-bellied sea snake, the greater the simian's chance of survival. A simple way to learn what's safe to approach and what isn't is by observing and copying the actions of others. It's much safer to let your cousin Francis determine whether that long yellow object has fangs. Over time you'll recognize the physical differences between the banana and the snake based on Francis's reaction to them, and you'll act accordingly if you find yourself in the same dilemma. Because you use the same neural pathways when observing an action as when performing that same action, you're more likely to bolt at the slightest indication of sea snake than to analyze the situation carefully. Over time, the "snake = danger" pathway reinforces itself in your brain until your reaction becomes an automatic behavior.

This heightened ability to mirror others' actions not only helps us avoid danger, but it also provides us with shortcuts to achievement. If you want to improve at a particular task, a step that is scientifically proven to be effective is to copy other people doing what it is you want to improve on or even simply thinking about excelling at the task. The scientists working with mirror neurons have provided an explanation of why this works.

Although this envisioning technique sounds like the power of positive thinking, it's not. Researchers have recently made distinctions between self motivation, by which participants tell themselves "Let's go!" or "I can do it!" compared to internal visualizations of completing a specific task. They found that pumping yourself up for a task by using motivational affirmations was *less* effective than mentally visualizing the process.

In one study led by Lien Pham at the University of California, Los Angeles, researchers asked students to spend a few moments each day thinking positively about getting a high grade on an exam. The students who participated obtained lower grades than those who focused on the process for doing well, including added study and review. The message is clear: Merely imagining achievement doesn't cut it. After reviewing numerous studies on the topic, I've concluded that visualization, as opposed to affirmation, can help by enhancing focus, increasing confidence, regulating effort, controlling cognitive and emotional reactions, and triggering automatic execution of a task.

By developing, exercising, and embedding mirror pathways, we're training our brains in a different way. For instance, if you want to be a better golfer, repeatedly visualizing every aspect of a shot—including observing other expert golfers practicing their craft—will improve your ability to play the sport. But you need to visualize in detail, in color, and using physical actions (e.g., swinging your arms as if swinging the club). What you're doing is training your brain as you would any other voluntary muscle in your body. Dancers do this every time they learn a new piece of choreography. That's why dance studios always have mirrors in them.

Our brains have far more processing power than we can ever comprehend, and most of this phenomenal ability occurs below the conscious level. We have to leave it to our brains if we want to excel. To understand what our brains can do once we provide them with a motive to take action, let's return to the example of hitting a golf ball.

Visualize Success

Here's a drill to train your brain to be better at a particular task: Begin by watching others do the same task (either on film or, better still, in real life). Observe people who are good at the task.

Next, imagine that the person you're watching is actually you. Now you're watching a film of yourself doing the task well. Remember to disassociate yourself from what you're watching. In other words, you're watching you *as an out-sider*. Now step into the movie and *become* you. Instead of watching yourself as an onlooker, move inside yourself and look out through your own eyes.

Continue to practice the task. For each new practice session, start by disassociating yourself and then move into an associated mindset. This will improve your brain and your ability to conduct the task.

Most amateur club golfers can hit a teed ball three hundred yards down the middle of a thirty-yard-wide fairway with some degree of consistency, but first-timers can't. Let's break down the shot. First the brain needs to work with the eyes to identify the direction in which to hit the ball. Then the body needs to align itself correctly. The golfer rotates his or her arms in a back-and-forth arc to swing the club—a four-foot stick with a hitting area no larger than a deck of cards—so that it hits the ball—which is barely more than an inch in diameter—at a speed in excess of one hundred miles per hour. This process involves input from more than forty muscle groups, and it might seem impossible that the brain could compute and carry out this task in mere seconds, but it does. Believe that your brain can do wondrous things, and it will.

We learn both by personal practice and by watching others, so you can shorten the learning process by observing people in real life, which is more effective than watching a video. The best vantage point is always the closest. The closer you are, the more detail your senses absorb and the more information your brain can use to improve the practice. Find an observation point that allows you to use all your senses. If you want to take up ballroom dancing, get yourself into a ballroom. If you want to improve your football game, study a training session from the sideline rather than in the stands or on TV.

On a parallel but related note, we constantly react and respond to what's unfolding around us. In fact, we're incapable of ignoring our surroundings, as we'll discover later on. If you absorb lots of cynical, negative television news, for example, then over time your mirror neurons will organize themselves in pathways not conducive to well-being and personal achievement. Eventually you'll trigger negative, cynical thoughts simply by seeing or thinking about your television set. When that happens, it's time to change channels and watch something nicer.

REMEMBER . . .

* Self motivation—psyching yourself up—isn't enough and ultimately doesn't work.
* To improve at a task, observe experts in real life and up close. Then visualize yourself successfully copying their actions.
* Regularly imagine the feeling of performing a task expertly.

{7}

Single Scripts and "Chunking"

YOU can visualize all you want, but you can't learn a complex skill instantly. Your brain has to break down incoming information into manageable, bite-sized "packets" in order to embed them into your long-term and muscle memory. The number of steps depends on a number of factors, including intelligence, interest in subject matter, and learning environment (learning in a dance studio versus on a bus, for example). The learning ability of a person or group is known as the cognitive load threshold (CLT).

Ruth Clark, Frank Nguyen, and John Sweller define cognitive load theory in *Efficiency in Learning* as a "universal set of learning principles resulting in an efficient instructional environment as a consequence of leveraging human cognitive learning processes." In other words, the CLT system explains how to make complex subjects easier to understand. The system clarifies how people trying to learn something process incoming information to retain it for future use. Put even more simply, it's how people remember.

Have you ever thought about how our brains can manage the most complex tasks with seemingly little or no effort? How can we cope mentally with all the functions involved in driving a car, for example? So many buttons, dials, knobs, levers, and pedals control a growling two thousand–pound beast traveling at breakneck speed and into some seriously small spaces. Our brains can perform these complex actions, and many others, by relying on single scripts and chunking.

Single-scripted behaviors are a sequence of expected behaviors in response to a given situation. Single scripts can be learned and, with practice, become part of larger scripts. Here's how a single script can help you improve your public speaking and presentation skills: Speaking in public can be nerve-racking (that old insula getting in the way again). But if each note card or slide is a trigger for a single script, the ordeal becomes more bearable. The key is to turn each slide, page, or card into something you know by heart. Rehearse the content until you know it inside out and upside down, and create a trigger from the title of the card or slide. By doing so, you're assigning all the details to your long-term memory (more about that later) so that your PFC only has to process the trigger you created.

People constantly follow established scripts and learn new ones. Habit, practice, and simple routine create new scripts. Forming and following them saves the time and mental effort of figuring out an appropriate behavior each time a new situation occurs. Examples include shaking someone's hand, "air kissing" business associates of the opposite sex, and giving a loved one a hug. Of course, scripted behaviors can be negative as well. For example, after giving up smoking, some people find themselves eating or drinking more. They've replaced the scripted behavior of using their hands to smoke every thirty minutes with another action, such as snacking or swigging a drink.

Learning a group of behaviors or actions allows us to carry them out subconsciously in sequence following a single stimulus. Let's take the example of the handshake. The initial conscious process is memory-intensive and includes a long list of actions and evaluations: How far should you extend your arm? At what angle should you hold your hand? How much pressure should you exert? How many shakes is too many? Once we learn the handshake script, we don't have to go through that complicated mental process every time we meet someone. Over time, our brains learn the handshake to such a proficient level that the sight of somebody walking toward us in the correct business context automatically triggers the handshake script.

My own research tested the level to which the handshake is an automatically controlled set of behaviors. In a small 2010 study, participants approached me, and I held out my hand. Just as our hands were about to interlock, I grabbed the hand of the participant with my left hand, said a number aloud, and then let him continue with the handshake with my right hand. Only one of the twenty participants noticed this incongruous behavior. The remaining nineteen admitted no knowledge of it happening, even though seventeen of them, when asked for the first number that came into their heads, repeated the number I'd said when grasping their left hands. I had interrupted a single-scripted set of behaviors, but, because I allowed it to carry on, the participants' brains didn't perceive the interruption, which allowed me to embed the number in their minds.

Think of the opportunities that these studies on subliminal messaging present to advertisers seeking to familiarize viewers with a brand, to business leaders intent on motivating employees, to teachers trying to connect with students. The list is endless.

———

The second prefrontal cortex assistance process that we use is called chunking. This technique involves creating a strategy for making more efficient use of your limited reasoning power by grouping smaller bits of information into larger chunks. It's a similar process to learning Morse code. The operator begins by learning each letter's dots and dashes. As the trainee grows more proficient, he chunks groups of dots and dashes into words. Then, as he becomes even more experienced, he learns to chunk entire phrases into manageable blocks.

Think back to the when you learned to drive a car. The first time you sat behind the wheel the process of driving was alien. For a beginner, driving a car is a mentally draining series of conscious evaluations and adjustments. New drivers must evaluate any number of situations constantly. As a result, they soon become cognitively "full." But as time passes, driving becomes easier until it's second nature. That's because we chunk groups of behaviors together to reduce the input needed from short-term working memory. What new drivers find so mentally taxing and stressful we can handle with ease once we've been driving for a while. Not only do we more experienced motorists drive more easily, but we can also drive while multi-tasking: holding a conversation, programming the radio, or finding and drinking from a bottle of water. (Some people stretch chunking too far, though, by believing that they can safely shave, put on makeup, or even change their clothes while driving, which is dangerous, not to mention illegal.)

Piano players learn to chunk without even realizing it. Students begin by learning individual notes—A to G—in each scale. Once they memorize the notes, they can begin to understand and learn chords. Pressing the C, E, and G keys simultaneously produces a C-major chord. Over time, players learn more chord combinations and then entire pieces of music. But they start with a basic, single script, and chunk it together with others.

The script allows the mind to retrieve these chunks from memory much more quickly and efficiently than if it were retrieving individual pieces of information. For another everyday example of chunking, consider memorizing a telephone number. It may be difficult for most people to memorize 4355558426. But if the digits are divided into three simple chunks (435-555-8426), it's far easier to remember.

How to Break a Habit

We all rely daily on single-scripted behaviors and chunking. The cigarette smoker who lights up twenty times a day has developed a single-scripted behavior to get his nicotine fix. His script entails reaching for the pack, retrieving a cigarette and lighter, lighting, and inhaling. The same person also has triggers that initiate each script. These may include making coffee, starting the car, or a host of otherwise innocuous actions.

Our goal in understanding scripting is to distinguish positive from negative actions. If, for instance, you want to quit smoking, what behavioral triggers cause you to light up? Tension? Boredom? Frustration? Alcohol? Once you identify a trigger, work on replacing it with something that leads to a different action. In other words, give your brain another task to take its focus off the trigger and subsequent negative behavior.

Let's say you enjoy a mid-morning cup of coffee along with a smoke. The trigger for the cigarette might be making the coffee, the smell of it, or even pressing the buttons on the vending machine at work. In this instance, you need to develop a replacement script that takes your mind off smoking. For example, drink your coffee in a place where you can't smoke, such as at your desk or in a public area. Or remove the coffee altogether and make yourself a smoothie instead. Either way, you're giving your brain another activity on which to focus at the exact moment it expects to prompt you to light up. Better still, take yourself away from the scenario in which you experience the trigger. If you smoke in the car on the way to work, don't drive, and take the bus instead. Whatever you choose, the aim is to interrupt the single script wired to the negative behavior that you want to change.

Some of us chunk recipes and then cook a meal with no further need for the recipe. An image of lasagna triggers all the ingredients needed, and if you're in a familiar grocery store, you buy them on a single pass down the aisles without backtracking. (This is why menu cards are such popular giveaways in supermarkets.) We can use chunking to achieve greater success, whether in the exam room, at work, or on the dance floor. If you try to recite the lyrics to a pop song cold, you probably have trouble doing it. But if you sing along to it on the radio, the parts you couldn't remember spring to mind once you hear the first few words. This is chunking in action, and it can be a powerful tool for personal achievement. What have you been putting off learning or doing because it's too overwhelming? Chunk it into smaller steps, and see how much easier it is to make progress.

Fear of the unknown severely restricts us from achieving our full potential. Too many of us avoid situations we haven't memorized instead of finding more efficient ways of memorizing them. Chunking offers you a powerful tool to learn and retain more mental stuff. Here's an example: A friend of mine took a job as an office receptionist. At first, she found it both counterintuitive and time-consuming to locate the right extension from a long list of names and numbers. The list was alphabetical by first name, so looking for Bill Smith meant searching through all the Bills and Williams in the company before finding Smith. The chunking solution was simple. We recreated the list, alphabetizing by last name and placing the corresponding first names afterward. So the entry became "Smith, William." Suddenly my friend could find extension numbers much more quickly, in a system that used more intuitive chunking.

REMEMBER . . .

* Identify your own negative internal scripts and interrupt them.
* Break daunting tasks into smaller, more manageable chunks.

{8}

Memory

WE have two fundamentally different types of memory: short-term, or working memory, and long-term memory. This binary is how evolution made the human brain more fit for its purpose. As our brains struggle to cope with everything that life throws at them, they've had to adapt and compromise—often crudely and inefficiently.

Many thousands of years ago, either up in the trees or balancing on hind legs in the savannahs of Africa, our human lives were simpler: kill or be killed, procreate, gather nuts and berries. Consequently, early humans had little need for memory. As time passed, however, we developed language, numbers, scientific principles, and many other abstract concepts, all actively contributing to our advancement. Each of these processes, functions, and activities takes up space in the brain, and—although this is somewhat an oversimplification—all the stuff that people need to remember must be accessible by way of a filing system to optimize memory efficiently. To quote Homer Simpson, "Every time I learn something new, it pushes some old stuff out of my brain!"

Faced with life-and-death decisions to be made in the blink of an eye, we don't want to have to root around in the archives of our minds for what to do. Conversely, the words to "Tik Tok" by Ke$ha shouldn't hover at the forefront of our thoughts every waking hour.

The different kinds of memory in the human brain resemble the different sorts of memory contained in modern computers. Short-term, or random access memory (RAM), handles basic functions. It's relatively limited in size and restricted in terms of how much information it can hold at any one time. Long-term memory is a much larger capacity hard drive that users fill with music, photos, data, and all they've learned in life so far. In both cases, RAM is supported by what's called virtual memory. This actually comes from the main hard drive to help with software and some hardware functionality, and it prevents the RAM from becoming overloaded.

Another analogy that illustrates how our brains function is to think of them as a social-networking site in which each member of the network represents a neuron. When one user sends a message to another, a connection is made. The more connections a user has, the more connections he can make, and a well-connected member of the network can communicate to all his connections in a split second.

Further, the more often the network makes the same connection between certain users, the more "hardwired" that connection becomes. Those we see as favorites or close friends on social networks are analogous to habits when it comes to human brains, and the brain manages these habits with little or no awareness or cognitive input. In other words, they're on autopilot.

Combining the computer memory analogy with the social networking analogy explains why we have two types of memory and how they interlink. We store all of our social network contact details in our long-term memories. As long as these contacts are online, the neural pathways are available to make contact with them. The short-term memory retrieves the minimum amount of details from the long-term memory to enable contact. Then it assists in the actual connection, which, once made, it leaves alone. As soon as your short-term memory has finished its involvement in a particular task, it erases all the data it used and moves on to its next pending task. The two analogies aren't perfect, but they do nicely summarize the two types of human memory functions working together.

Our short-term memories handle the mundane, minute-to-minute, limited-time-span decisions we make. Oxford professor Susan Greenfield gives detailed evidence of this part of the mind's limited functionality in her excellent book *The Private Life of the Brain*, explaining that human short-term memory can store information for no longer than eighteen seconds. This explains why so many of us can't remember why we walked into a particular room or how we behaved a minute earlier.

Fortunately, we don't rely solely on our short-term memories. Our long-term memories, which are more powerful processing units, store everything we've experienced. To give an example of its phenomenal power, just seeing or hearing the name of a particular location can trigger the memory of a vacation from years past, including where you stayed, what you ate or drank, what the weather was like, who you were with, and an incredible amount of additional details. Unfortunately, the long-term memory also stores a lot of bad memories. These normally are locked away, but they can jump out when least expected and screw up your day. When that happens, you need to dismantle the bad memory and replace it with a success anchor, which we'll learn how to do a little later on.

So far we have looked at long-term memory as a single entity, but it actually operates through the comingling of three different types of storage. The first is procedural memory. We use this for learned skills such as playing a musical instrument. To embed something into procedural long-term memory, we have to learn the procedure, often from many hours of laborious practice. It takes many repeated attempts at the new procedure before the final coordinated execution—involving brain, senses, and motor skills—can play out. This form of manually hardwiring the brain can be difficult because our brains tend to forget things until they embed them. As a result, we have to keep relearning parts of a procedure until the entire procedure is chunked and scripted.

The next long-term memory type is declarative memory. This kind is most active at school when we're learning a foreign language or mathematical concept. Declarative memory is used in the recollection of facts, figures, and pure knowledge. Humans never stop learning, and as the saying goes, the more we know, the more we grow. For those of us committed to success, it's advantageous to recognize the value of what we're trying to learn. Think of learning as retaining information. The more we retain, the more we know, and the more we know the more we grow.

The third type, which many consider similar to declarative memory, is episodic memory. As the name suggests, this kind is responsible for storing discrete installments or episodes from life. Remembering a trip to the coast or an evening out essentially is reliving a past event or episode. Episodic memory retains information relating to larger episodes much more easily than with either procedural or declarative memory. According to Trygg Engen, "Memories triggered by episodic odors fade only by 3 percent over a 12-month period," which explains how the smell of the ocean or the scent of humid air can bring to mind an entire summer, long past.

All of us find it easier to remember episodes (good and bad) than procedures or languages, for example. So remembering a procedure as an episode will make it stick in your memory more easily. If you want to remember how to use a particular computer application, turn it into a story or an episode. The more emotion you include, the more effectively you'll memorize it. For example, instead of thinking about a spreadsheet full of data, create a story that contains each of the columns. Sultry maiden ColumnA threw a rope to ColumnB, who knotted it so the two sets of data were always next to each other.

All three forms of long-term memory use a clever way of indexing their information. They store it as components that form meaningful content when associated with the right criteria and within the appropriate context. For example, directly in front of you is a metal box with rounded edges. It's white and has lights on the front. Those two sentences could describe a large number of objects, but as the brain receives more details our long-term memory filters out some of the options. With more context, the object becomes clearer: It has a round door made of translucent glass, and the white box is in the laundry room. Finally we understand that the box is either a washing machine or a dryer.

In principle, that's the way long-term memory operates. By using this technique of filtering and association, the brain can use stored components as parts of different memories, and that system aids storage efficiency and recall. An important means of embedding information into our long-term memories—particularly episodic—is to attach emotions and feelings to them. That's exactly what advertisers often do. They take an everyday product, such as a bottle of disinfectant, and attach an emotional association—a mother caring for her adorable toddler, for example. Suddenly, the disinfectant goes into the "caring" section of our long-term memory.

REMEMBER . . .

* Short-term memory forgets; long-term memory remembers.
* The three kinds of long-term memory are procedural (practicing a skill), declarative (facts and figures), and episodic (installments from life).
* It's easier to remember episodes than procedures, so if you need to remember how to do something, turn it into a story.
* Emotion reinforces long-term memory.

{9}

The Sensory System

OUR brains receive signals from our bodies in the form of sensory stimuli, either at a conscious or unconscious level, by way of the central nervous system. These stimuli enter the picture via one or more of our five senses: sight, sound, touch, taste, and smell, which scientists correspondingly describe as visual, auditory, kinesthetic, gustatory, and olfactory. Our senses tell our brains about the environment around us: what surrounds us, where it is, and other salient details that our brains might need to ensure our survival.

One example of how our brains and senses work in tandem is the speed with which we recognize our friends. Take a look at a photo, watch a video, or even scan a crowd. Your eyes immediately send information to your brain, which uses its billions of neurons and mental pathways to process the image. It takes less than half a second for the brain to determine whether a facial image in a photograph is someone familiar.

Each of us can recognize thousands of shades of color and differentiate among thousands of smells and textures. We can tell the difference between being tickled with a feather and poked with a stick. But if our senses are so amazingly effective at communicating data to our brains, why do we need to learn more about the process?

The brain continually makes educated guesses based on the information it already has and also on the environment around us. For example, we see shapes in clouds during the day and faces in trees and noises that sound like footsteps at night. In a heightened state of awareness, our brains are trying to find recognizable patterns in order to help us survive. When they have minimal amounts of sensory data, they keep us alert by guessing.

Here's a simple tool you can use to "deceive" your brain into delivering the personal achievement you crave. First, remember that your eyes are constantly looking for possible dangers. Although you focus only on a small area in front of you, your eyes are communicating much more to your brain than what's in that spot. If something unexpected moves in your peripheral vision, you'll turn to focus on it. That means that your brain is reviewing the entire vista around you. So if you're staring at an image of something you wish to own as part of your visualization success strategy, pay attention to its physical presentation. What else is in the same line of sight? Let's say you have a photo of a Mercedes Benz on your desk. That's an appealing image but less so if it's next to an unpaid stack of bills. So while the aspirational image of the car will help charge your brain, seeing all those bills at the same time will dash your hopes. In the example of sitting at your desk, make sure everything in view is positive and all the negative triggers stay out of sight and therefore out of mind. (But don't forget to pay your bills!)

Here's another example that a lot of people use often. Our brains adjust our metabolic rate in relation to music. Scientists studied the effect both of soothing and exciting music on the respiratory functions in people, concentrating on oxygen consumption and metabolic rate. Soothing music decreased oxygen consumption and metabolic rate, while exciting music increased them. If you need to feel more energetic, play faster music. Conversely, in times of stress, reach for your "chill out" playlist.

As a final example, the cost of an item influences how we think it tastes. Numerous studies have shown that, even if wine A and wine B are the same, people prefer wine A if it costs more. In other words, people—particularly those sneaky marketing and advertising execs—are constantly deceiving our brains.

We know that our senses don't provide the full story to the brain, so it needs to fill in the blanks and guess at the rest. That's the bad news. The good news is that knowing this gives you the opportunity to feed information to your brain that allows you to shift your brain to be more in line with your success aspirations. If, for instance, you're self-conscious and think that others are looking at you, chances are your brain will send blood to where it thinks people are focusing their attention—typically, your face. The result? You blush. But shift your mental focus, maybe to your new high-end wristwatch or shoes, and you'll reduce your self-consciousness along with the reddening in your cheeks.

Our senses, in partnership with our brains, can distort our perception of reality, so making sense of our senses makes . . . well, *sense*. Let's take a closer look at how the five primary senses work both for and against us.

SIGHT

Vision, one of the most important senses for humans, requires input from the eyes that the brain then processes. But, as you know if you've ever played a hidden-object game, looking and seeing are two totally different processes. We'll learn about them shortly. Meanwhile, here's a basic explanation of how human sight works.

Light enters the eye through the cornea, passes through the lens, and hits the retina at the back of the eye. The retina contains two types of photo-receptor cells called rods and cones. These send corresponding impulses to a network of neurons, which in turn generate electrical impulses that the brain analyzes and decodes into what people are viewing.

In addition to two types of photoreceptors, we also possess two different types of vision: foveal and parafoveal. Foveal vision refers to what we look at with the center of our eyes. A small area at the back of the eye, the fovea comprises less than 1 percent of retinal size but takes up more than 50 percent of the visual cortex in the brain. Foveal sight governs our sharp, central vision, roughly a twenty-degree field of vision straight ahead. We use it for reading, watching TV, and everything else that requires focusing on detail. Driving a car also uses foveal activity—but only partly. That's because the brain is also monitoring what happens in our peripheral vision, which includes movement in the side and rearview mirrors.

The second type of vision, parafoveal, takes in general vistas and what's going on in the periphery. Over the course of evolution, we humans had to keep watch for predators, meals, and mates, which weren't always close at hand. It follows, then, that our surroundings influence us more than we might be conscious of. Researchers, such as cognitive psychologist Manuel Calvo at the University of La Laguna in Tenerife, Spain, contend that emotional images have more impact if targeted toward parafoveal vision. If the inside of your car is a mess, full of food wrappers, cups and cans, and other debris, the state of your car could influence your state of mind and affect how you feel. The same goes for how you dress. For you to attain maximum personal achievement, you need to surround yourself with the right visual stimuli in both your foveal and parafoveal vision.

Most visual stimuli attract our attention by entering the parafoveal vision first. If there's nothing sufficiently stimulating in the foveal part of your sightline, then what you're seeing is being processed entirely in the background. From the point of view of success, your eyes are sending information to your brain about much more than the specific objects on which you're focusing visually. They constantly monitor your overall vista so they can initiate a course of action should the need arise. This process directly impacts how you feel from moment to moment. Success doesn't depend only on looking at a photograph of a Mercedes Benz on your desk; it also depends on the entire desk, office, and even the view outside your window. Your brain is keeping an eye on all of those things.

Some parts of the brain are better at processing incoming visual data than others. These efficiencies or inefficiencies create the difference between looking and seeing. We can't process more than 5 percent of the visual information coming from the millions of rods and cones in our retinas, so we must extract information from the most meaningful patterns such as hard edges, contrast, brightness, and so on. If something catches your eye, that means you've looked at it, but only if you then process it have you seen it. Think about one of those hidden-object games where the instructions tell you to find the hidden tiger or the face in the photo. Your eyes scan the whole image, looking at all the changes in shape and color, and then when you do spot the face or the tiger, even if you look away and back again, you can't unsee it. Your brain has processed and absorbed that image.

As you can imagine, you look at much more than you see, and much of what surrounds you influences how you feel from moment to moment.

SOUND

Sound is a periodic compression of air, water, or other medium. It consists of waves that vary in length and height, which affect a sound's volume and frequency. The ear detects these waves, initially by the tympanic membrane, then with other structures, and converts the sounds into electrical impulses that travel through the limbic system for processing. The limbic system is a network of brain structures that govern instinct and mood. It controls the basic emotions (fear, pleasure, and anger). An involuntary emotional reaction takes place as the impulses travel to and are processed by the brain's secondary auditory receptors. The precise emotional reaction depends on how the sound is perceived (threat, meal, or mate).

James Kalat, author of *Biological Psychology*, holds that although people spend much of their time listening to language, they sometimes forget that the primary function of hearing concerns simpler but more important issues. What's making that noise? Where is it? Is it coming closer? Is it going to eat, feed, or mate with me? The human auditory system is well suited to resolving these questions because our brains are constantly listening. Our ears have evolved to catch sound waves effectively, which our brains then process as data. Often, we're more influenced by what we hear than we'd care to admit. Casinos tune the sounds that their gaming machines make to the key of C so that dissonant noises don't drive you away from the one-armed bandits. Carmakers design car doors so they make a solid *thunk* when closed, because manufacturers learned long ago that that sound communicates a sense of "quality construction," making potential customers think that the car is solid and well made.

But what does all this mean in your drive for personal achievement? Remember that our brains are monitoring the sounds around us all the time. Do you have a radio or TV station on in the background? Are you tuning into something successful or something more negative like the nightly news? Certain songs always inspire you, so if you have an MP3 player and can create playlists, make one called "Success" and play it often. It will help remind different parts of your brain that you're going to achieve more.

Soundtrack for Success

You can refine the sounds around you to enhance your mental representation of your individual achievements in other ways. For example, the sound of laughter will make you happier, so tune in to something funny on the way to work instead of the weather and traffic report. When you're out walking, ask yourself what your footsteps sound like. Are they a clear, up-tempo beat or more of a slow, lethargic shuffle? Remember that you are what you think you are, and incoming stimuli—even stimuli that you generate yourself—play a part in molding a successful you.

If you're traveling to an important job interview, what you hear along the way will alter how you feel. If you want to appear masterful and confident, listen to winning music—Tchaikovsky's "1812 Overture," John Philip Sousa's "The Washington Post" march, Queen's "We Are the Champions," or Survivor's "Eye of the Tiger" for example—which will literally get your pulse pounding. If you invite that hot new hire in accounts receivable over for dinner, play the right music in the background to set the mood accordingly.

The language others use can influence how you feel, but you can use that information to your advantage. By training your brain to have a clear notion of what each aspect of your success sounds like, you can reinforce your success playlist by playing it on your computer or smartphone. By doing this, you're providing your brain with yet another set of effective stimuli and clear instructions about what you want it to deliver. Sounds of success may include a round of applause you received after delivering a presentation or a particular music track that was playing when you achieved a particular significant goal. A few other examples: Someone else's tone of voice can have a dramatic impact on how you feel. "I love you" whispered in your ear elicits far different feelings than someone yelling the phrase at you, and the sound of uncorking a wine bottle elicits a greater sense of quality than the feeling of unscrewing a metal cap.

Work at recognizing not only the foreground sounds in your life but also the backgrounders. Are they congruent to your success or counterproductive? Does that constantly quarrelling couple in the apartment down the hall sap your energy and bring you down? If so, try to drown them out with more pleasant sounds. If you can't, then reframe any negative sounds you hear. For example, if you hear them screaming, congratulate yourself aloud on having a brain that would never get *you* into such an antagonizing relationship! Your brain will keep you on track toward your personal goals and achievements.

Just as with vision, hearing involves noises all around us, much of which we filter out and some of which we process mentally. As you develop a greater awareness of what's going on around and inside you, you'll recognize sounds that have a positive impact on your emotional state of mind. When you determine what they are, either record or download them and add them to your success playlist. When you identify noises adverse to success— sounds that make you feel lethargic, negative, or unmotivated—isolate and remove or reframe them.

The Ancient Greek philosopher Epictetus once said, "We have two ears and one mouth so that we can listen twice as much as we speak." Unfortunately not all of us abide by that maxim. When you engage in conversation, listen to yourself as well as to others. How do you sound to other people? It just may be that you've developed some bad verbal habits ("The problem is . . ." or "That's true, *but* . . .").

With this information in your toolkit, you have even more power to become the person others want to emulate. You can develop that power by controlling not only what you hear but also what others hear from you.

TOUCH

When you enter a room in complete darkness, your first action is to fumble around for the light. As your hands feel the wall, seeking the familiar shape of a light switch, your brain is creating a mental representation of what you're contacting. You're "seeing" without using your eyes. You do this by using the third human sense and the only one not isolated to the head: touch.

Skin—considered the largest organ of the human body, accounting for about 15 percent of total body weight—constantly receives and sends touch-related stimuli to our brains. The sense of touch relays a number of aspects, including shape, pressure, weight, cold, warmth, and texture. Skin has many different types of receptors that transmit different touch sensations.

Most of us know that tactile sensation conveys information about the object or person we're touching. But how do our brains interpret that information, and what courses of action might we take as a result? Once again, science provides a number of key answers. Researchers have concluded that an object's texture, hardness, and weight influence our perceptions, judgments, and decision-making process. In other words, how objects in contact with us feel can influence everything from the way we interact with others to the brand of smartphone we buy.

At a subconscious level, we can make irrational quality judgments by linking touch with a particular environment or situation. For example, the feeling of a car's steering wheel leads our brains to evaluate the likely quality of that vehicle. The same goes for the weight of a digital camera or mobile phone: too light and our brains think they're cheaply made.

But we can feel across our entire bodies. What you sit on influences what you perceive and think as much as what you handle. Give some thought to what the physical feeling of success means. Is it the leather steering wheel of that Mercedes Benz? If so, what does it feel like? Create a positive detailed mental representation of that feeling, and analyze it. Does it make you smile, confident, proud, satisfied? If success means losing weight, ask yourself how you will feel when you've finally shed those pounds. Do you want the clothes you own to fit more loosely, or do you want to buy new form-hugging outfits? Create the most vivid mental representation you can. Don't forget to tell your brain that this is how you *will* feel once it happens, not merely just how you *want* to feel.

Another technique involves the humble pen. Most of us write as part of our everyday lives, and one way to signal quality, seriousness, and intent to your brain is writing with a quality pen. Good pens are typically heavier and fatter than run-of-the-mill ballpoints. If you don't have one, get hold of one—literally. Writing with a heavier pen will help your brain attribute more heft to what you're noting.

What you're doing in each of these cases—car, clothes, pen—is duping your brain by using the sense of touch. The ability of touch to affect choice has its limits, but in the right circumstances it can be dramatic. Want proof? When Kathleen Parker of the *Washington Post* handed out clipboards with a job applicant's résumé attached, the staff who received heavier clipboards gave the job applicant higher ratings, deducing *physically* that those applicants were more qualified for the job.

Tactile Tactics

Scientists have long known that people sitting on hard surfaces make firmer business decisions, while those sitting on cushioned seats tend to have a softer touch. When you go on a job interview, note what kind of chair the interviewer is sitting in and ask yourself what your chair feels like. If it's not conducive to professionalism, expertise, and authority, reframe your impressions to reorient your brain to something more positive and goal-oriented. Instead of thinking about that cheap chair and chintzy carpet, concentrate on that big expensive-looking desk or perhaps the soothing view from the window. Then give your brain the icing on the cake it wants by pulling that heavy pen from your designer handbag or briefcase to take notes. Remember, you're in charge.

When deciding where to go on a date, you may want to consider your seating. Do you want your paramour to be a tough negotiator or more compatible? Don't overlook the other objects you both will touch on the date. The type and weight of cutlery in the restaurant and the carpet on the floor of the hotel lobby can influence the ultimate success—or failure—of your rendezvous.

Touch is as likely as all of our other senses to influence our state of mind and likelihood of achievement. What we touch does influence how we feel, act, and more. Give your brain all the success-oriented information it needs to propel you to your goals. The more you communicate achievement to yourself, the greater the success your brain will attain for you.

TASTE

Taste is defined as the stimulation of the receptors that cover the tongue, which we call taste buds. But James Kalat clarifies that when referring to the taste of food people often mean its flavor, which is the combination of taste *and* smell.

Of course, as with our other senses, our brains can misinterpret taste. Remember the research that revealed that people preferred the wine they were told was expensive wine over the wine they were told was cheaper, even though it was all the same wine? More than just what hits our taste buds can influence our sense of taste. It works the other way around, too. What a dinner date eats will color his or her perception of you, whether it's the lobster risotto or a jumbo chili dog.

As we know, our brains send and receive information constantly, so it's up to us to act as the executive chef to refine that information and align it with our personal achievement goals.

SMELL

An old First Nations Native American saying goes, "When a pine needle falls in the forest, the eagle sees it; the deer hears it; and the bear smells it." Unfortunately for us humans, smell is the weakest of the five senses and weaker by far in comparison with most animals, bears, birds, and deer included. We have about forty million olfactory receptors that detect up to ten thousand different odors. That might sounds like a lot, but it's barely scratching (or sniffing) the surface.

We can recognize smells from the moment we're born, concentrating first on odor-related danger signals, such as sour milk or rotten food. Our olfactory systems, wired directly to the part of the brain that contains the pleasure and survival centers, connect with the older parts of the brain that regulate emotion, including parts of the reptilian brain and the limbic system. As a result, we don't often rationalize what we smell, but we do have an immediate reaction and a subsequent tendency to act accordingly.

Olfactory stimuli don't pass though the brain's mental filtering system, so we tend to evaluate smells purely at face value. In that respect, context becomes king. Often we must rely on other senses to confirm what a smell actually is. Our noses alone might be unable to tell a ripe Camembert cheese from a pair of smelly sweat socks, but in the context of a high-end grocery store or a locker room, the meaning of the smell becomes more obvious (hopefully!).

When it comes to how we smell things, a significant difference exists between men and women. On average, women detect odors more easily than their male counterparts, and the female brain's responses to them are stronger, too. Women are also more likely than men to care about the smell of a potential romantic partner—no surprise there.

As with our other senses, we're constantly receiving smell-related sensory information. Unfortunately our poorly wired noses quickly become accustomed to the smells around us. For that reason, not everyone realizes when he or she has a personal hygiene issue. But each of us can take steps to manage how we smell to others: regular grooming, good dental hygiene, and clean clothes, all of which also make us feel good.

As with your other senses, create a smell-related representation of how a successful you—and how success itself—should smell. Perhaps you noticed a particular brand of perfume or aftershave that you associated with success when you met someone powerful or serene. Maybe your smell of success comes from the freshly laundered and starched sheets in a five-star hotel. Whatever success smells like to you, develop a mental representation of it and then imagine that representation as a reality.

An example from my own life: Early in my business career I realized that a number of successful men I knew had a distinctive aroma. In time, I learned that they all wore the same deodorant, Chanel Allure. I associated that smell with their success and have worn the same product every day since.

A word of warning though! Once you think you've identified the ideal smell for yourself, ask a trusted friend for his or her opinion. Remember, our sense of smell quickly acclimates, so we lose track of something even if it's unpleasant. Getting a second and unbiased opinion never hurts!

———

Now that we've explored how each of our senses works and the ways our brains interpret all that incoming sensory stimuli, let's sleep on it.

REMEMBER . . .

* Pay attention to the context and physical presentation of objects that define success for you. Don't let accidental negativity surround them.

* If you blush when embarrassed, redirect your attention to something you like or that makes you proud, and the redness will fade.

* How you dress influences your state of mind and how you feel, so dress for success to achieve it.

* If you need more energy, listen to up-tempo music. When you need to relax and decompress, listen to calming sounds. Make a success playlist of inspiring and stimulating songs, and play it often.

* Start your day right by listening to the sound of laughter to put you in a happy mood.

* If you can't avoid negative noises, make a point of reframing them mentally.

* What does success feel like physically? Create a mental representation of it, and study it to find out what you want in life and what makes you happy.

* Write with a heavy pen to imbue your words with greater meaning.

* If you want to have the upper hand in business negotiations, sit on a hard chair or surface and make sure others sit on softer surfaces.

* Determine your smell of success and make it present in your life—but remember to ask a trusted friend for his or her opinion about it.

{10}

The Science of Sleep

YOUR brain is an incredible organ, and like most organs it needs regular periods of rest. The average person spends around eight hours a night sleeping. That works out to fifty-six hours a week, just over nine days a month, or about four months of the year asleep. Our eyes are shut, we don't respond to sound or light, and our bodies relax. But just because we spend a third of our lives dozing doesn't mean our brains shut down. In fact, our minds continue to work around the clock even while we snooze. But doing what?

Using a technique known as electroencephalography, researchers have been able to record the brain activity of people who are sleep. They have identified two different forms of sleep: rapid eye movement (REM) and non-rapid eye movement (NREM) sleep. While we sleep, our brains cycle through those two kinds of sleep several times during a typical night. Most dreaming occurs during REM sleep, when our muscular activity is at a minimum. One possible reason for that is to prevent us from physically acting out our dreams—thankfully!

Sleep gives our bodies a chance to heal and recover from all the activities in which we engage while awake. Sleep also allows us to conserve energy. After having fought, fled, and fornicated our way through the day—once we've taken in enough sustenance (food and water) to last for a while— our bodies shut down many of our physical functions for a few hours. Researchers also believe that sleep, especially REM sleep, helps our brains embed memory and learning. In other words, sleep gives the mind the means to form memories and "file" them away where they belong.

In tests conducted over the past several years, medical researchers had people press buttons when they saw certain images on a computer screen. Their performance improved with practice, which makes sense. After all, practice makes perfect. More surprisingly, though, performance also improved in subjects who had had a good night's sleep. The researchers concluded that that measurable improvement signifies the importance of REM in memory processing.

Nearly everyone knows that it's important to get a good night's sleep the night before performing a strenuous mental task, such as taking an algebra test or giving a piano recital. But going to sleep immediately after studying for that task helps your mind sharpen that information or those skills you just learned. Remember, the brain at rest continues to improve memory and file information away for easier retrieval later. By studying, practicing, tuning, and fine-tuning what you need to learn and then going to sleep immediately afterward, you can wire that information more strongly into your memory. So sleep aids the formation of memory both before learning something new *and* after.

So if sleep helps our minds organize memory and process the information of the preceding day, can it help even more? Possibly so. Research shows that sleep may also help us solve our problems. A 2009 study by social psychologist Ap Dijksterhuis and others suggests that our unconscious minds might be hard at work even while our bodies are sound asleep. Unconscious thought is an active, goal-directed mental process, but in unconscious thought the usual perceptions, opinions, and attitudes that form our conscious thoughts are absent. In unconscious thought, we weigh the importance of each aspect that makes up our decision more equally, leaving our conscious biases within consciousness. In other words, our brains can make more objective decisions when we sleep, giving us one more way to harness the power we need for our personal achievements.

What researchers have discovered squares with what the age-old aphorism suggests we do before making a big decision: "sleep on it" before pursuing a course of action. The next day? *Voila!* Decision made, action planned, piece of cake. An opium-soaked dream brought Kubla Kahn and his stately pleasure dome to life for Romantic poet Samuel Taylor Coleridge. After a night of hearing and telling ghost stories, Mary Godwin (later Shelley) had a nightmare that became *Frankenstein*. Salvador Dali and his colleagues created the genre of Surrealism largely from their dreams. Marion Jones dreamt that she'd broken a world record and then brought the dream to life.

Here's an example of how you can do the same. Ask your brain a question, but let your brain come up with the best answer. That sounds easy enough, but remember that the brain doesn't always respond in the ways that common wisdom suggests. We first need to give our brain a way of coming up with the correct answer, and second, a way of telling us that answer *without* interference from the conscious part of the mind. So how can you accomplish such a Herculean task? Try this.

Exercise
Sleep on It

Write a problem on a notepad just before you turn off the lights and go to sleep so that your brain has a fresh task to tackle as you sleep. If you've already detailed the problem in writing earlier, read it again before getting into bed.

Here's the important part. When you wake the following morning, reach for your notepad, read the problem, and immediately write down the first thing that comes into your head. More often than not, that will be your unconscious mind's solution for you.

But remember, urgency is key. Why? Immediately after you wake, you can still tap into the creative right hemisphere of your brain (before the chatty left hemisphere gets going). Once your conscious mind wakes up, it distracts your subconscious and clouds its judgment. So within seconds you can lose the unconscious solution to the problem. But if you can beat your conscious mind to the punch by writing down your first thought of the day, you have a clear alternative solution to your problem.

This doesn't always work after a single night's sleep, though. Sometimes the brain might require two, three, or even more nights of unconscious wrangling with a problem before coming up with the best solution. You might need to repeat the process over a number of nights. But the wait is worth it. To help you achieve the success you want, you need to call on all the help your powerful brain can give you, even the less conventional.

REMEMBER . . .

* Our brains continue working even while we sleep.
* Sleep improves memory.
* Going to sleep immediately after learning a skill or new information embeds it in the mind.
* Sleeping on a problem overnight allows your brain to work on solving it.

Maintain Your Brain

A T a fundamental level, eating well is good for both your mental and physical health. Just as your body and muscles require nutrients, so does your brain. It's especially important to eat right for your mind because, on average, your brain uses 20 percent of the energy you consume while accounting for barely 2 percent of your body weight. The food and other substances you put into your body affect your behaviors, emotions, moods, and thoughts. Thankfully the brain responds quickly to proper nutrition. A good diet not only boosts the brain's abilities but also helps prevent the onset of diseases such as dementia.

Our brains require essential fatty acids (EFAs) to function properly. The two main EFAs to ingest are omega-3s and omega-6s. The brain particularly likes omega-3s, which help organs function properly and foster brain-cell activity by helping cell walls form and facilitating oxygen circulation. Omega-6s help the body fight skin diseases, cancer cells, and arthritis. Unfortunately the human body doesn't produce omega-3 and omega-6 fatty acids itself, so we must get them from the foods we eat. Good sources of these vital acids include oily fish (anchovies, herring, mackerel, salmon, sardines, trout, and tuna), various nuts (Brazil nuts and walnuts in particular), seeds (flax, hemp, pumpkin, and sesame seeds), plus avocados, soybeans, and dark leafy vegetables (kale, spinach, mustard greens, and collards).

The brain acquires its energy from glucose, a form of sugar transported throughout the body via the bloodstream. A sugary snack or drink rapidly raises your blood sugar level, but it can deliver too much too quickly. Because the neurons in the brain can't store glucose, they depend on the bloodstream to deliver a steady supply. For that reason, it's better to ingest complex carbohydrates, which act like slow-release capsules of sugar, instead of simple carbohydrates, which are more like a sudden shot. Complex carbohydrates generally are found in natural—meaning unprocessed—whole foods, and they have long chains of sugar molecules that the liver gradually breaks down into the shorter glucose molecules the brain uses as fuel. Simple carbohydrates, although present in some natural foods, are found mostly in processed or refined foods. You can find complex carbohydrates in wheat cereals, whole-grain bread, whole-grain pasta, and black and brown rice.

When processed into juices, pastes, or sauces, tomatoes contain high concentrations of the powerful antioxidant lycopene, which helps protect against damage to brain cells that can lead to forms of dementia. Broccoli contains vitamin K, which is good for enhancing brain power by keeping blood flowing well and removing damaging heavy metals.

Here's how to eat healthy for your brain throughout the day.

BREAKFAST

In the morning, most people consume a high-carbohydrate breakfast of cereal, toast, or croissants. The body craves that kind of food when blood sugar levels run low, but high-carb foods cause the release of serotonin, a brain chemical that makes you feel relaxed and satisfied. Serotonin is an essential neurotransmitter for a well-balanced mood, but it creates too relaxed a state right after breakfast. A breakfast full of complex proteins, such as eggs and yogurt, energizes the body, so those are better options.

A key messenger in the brain that keeps brain cells excited is acetylcholine, which is essential for concentration, focus, muscle coordination, and storage and recall of memory. You can find the choline part in eggs, liver, and soybeans.

Caffeinated coffee is a classic remedy for sleepiness, but drink too much and the pituitary gland interprets this spurt of chemical activity as a warning sign of impending danger. Your body then produces adrenaline, which makes you feel anxious. A single espresso can help with alertness, but a double or more may incite anxiety, so don't overdo it at the coffee pot.

LUNCH

As with breakfast, it's important to eat a lunch that keeps you energized throughout the afternoon. Simple carbohydrates—contained in many processed foods, sweetened chocolate, and soft drinks—spike serotonin levels in the blood but also put you to sleep. Proteins and complex carbohydrates (whole grains, fruits, vegetables, and nuts) give you energy and keep you motivated because they release glucose at a slower rate and provide a steadier supply of energy to the brain.

DINNER

The brain's needs change when it comes to the evening meal because some foods don't stimulate the release of hormones that promote wakefulness and focus. Because it consumes a disproportionate amount of the body's oxygen supply, the brain is more vulnerable to oxidation from free radicals. The brain needs vitamin E, which protects fatty molecules in the brain and helps it remain functional and healthy longer. A study conducted at the National Institute of Aging found that older folks who took vitamin E supplements were less likely to die prematurely than those who hadn't taken the supplements.

Dinnertime is when you should eat that salmon, packed with omega-3 fatty acids that the body can't produce. Spinach is a good vegetable to accompany the salmon because its antioxidant properties neutralize those harmful free radicals that drive tissue damage and aging. You also should eat brown rice and strawberries to boost your memory and levels of vitamin E.

We also need to ingest vitamin C, and doing so in the evenings feeds the brain as it embeds the day's learning into memory as you sleep. There is a good amount of vitamin C in bell peppers, berries, broccoli, dark leafy greens, goji fruit, kiwis, oranges, papayas, peas, and tomatoes. Another good food to eat later in the evening is cottage cheese, because it contains high levels of the amino acid tryptophan from which serotonin emanates and helps you fall asleep.

———

To optimize brainpower, Michael Green of Aston University suggests eating "more frequent but smaller meals. The brain works best with about 25 grams of glucose circulating in the bloodstream—about the amount found in a banana." He added that "People who eat diets high in saturated fat are more susceptible to cognitive deficits, including an increased likelihood of strokes. Rats that gorged on saturated fat for several weeks had obvious damage to the hippocampus, a part of the brain known to be critical to memory formation."

Don't forget to drink plenty of water throughout the day as well—and carbonated sodas, juice, coffee, and tea don't count, sorry! Brains don't perform well "dry." When parched, we have more difficulty focusing. Dehydration also can impair short-term memory function and the recall of long-term memory. Simplistically, the thirstier you are, the more your brain focuses on that thirst and the less it can attend to anything else. You want to satiate your thirst without ingesting other less healthy chemicals such as sugars and other preservatives found in processed drinks. Often a headache signals dehydration, so the water you drink with that headache medicine could be as beneficial as the pills . . . if not more so!

Everyone knows daily exercise is good for the body, but it's also just as good for your mind. Exercise feeds increased quantities of blood and oxygen to our craniums, and researchers believe that considerable health benefits accrue in the brain as we elevate our heartbeats. For example, taking a brisk walk has been shown to improve memory in the elderly. Running leads to increased brain-cell numbers in normal adult mice as well as in human brains. People who exercise regularly generally have better short-term memories and quicker reaction times. Cognitive psychologist Lorenza Colzato of Leiden University in the Netherlands found that those who exercised four times a week could think more creatively than those with a more sedentary lifestyle. Unfortunately, though, with a sedentary lifestyle comes less physical activity as part of daily life, so you have to make a concerted effort to exercise regularly.

Exercise can be mental as well as physical. Our brains develop by interacting with the world through perception, thought, and behavior. It's important to keep challenging your brain because mental activity stimulates the creation of new neurons throughout your life. The more children tax their brains, the greater their cognitive ability in adult life. This is otherwise known as education! Mental stimulation improves brain function and also fends off disease and age-related cognitive decline.

Even in old age, the brain can grow new neurons and develop new neural pathways between them. Scientists used to believe that once you reached adulthood your brain lost its ability to form new internal connections. They thought that ability, called plasticity, was confined to infancy and childhood. But a 2007 study found that a stroke victim's brain had adapted to damage to nerves that carried visual information by extracting similar information from other nerves. This followed several studies that showed that adult mice could form new neural connections. Later studies found more evidence of human neurons making new connections into adulthood. So it's never too late to rewire your brain, but it works both ways—you can program both good and bad connections.

Learning and memory involve changes at neuron junctions called synapses. Those changes make communication between neurons and memory formation more efficient. Neuroscientists at Brown University showed that learning produces changes in the synaptic connections between brain cells. After the researchers taught rats a new motor skill, they found that the animals' brains had changed. The strength of the synapses in the motor cortex had increased.

The trouble these days is that the more work that machines do for us, the less we have to do ourselves. The same holds true for our brains. As computers do more of life's mental lifting, we have fewer reasons to think, which of course isn't conducive to training our brains for success!

In the old days of mom-and-pop shops, you could select ten different items, and the cashier mentally tallied your total in mere seconds and it took the same amount of time to calculate your change from a twenty-dollar bill. Because today's register attendants rely on their machines to do the math, it now takes some of them much longer to do basic math with even small amounts . . . as you may have experienced in the middle of a transaction when paying with cash and finding some pocket change that you want to unload.

Think of your brain as a muscle. To keep it working in tip-top shape, ready to spur you onto every success you've ever desired, you need to exercise it. Find twenty or thirty minutes each day to tax your brain with something new, and remember, it really is the thought that counts when striving toward greater personal achievement.

Mental Workout

To give your brain a workout, Amir Soas of Case Western Reserve University Medical School suggests reading (you're already on the right track!), doing crosswords, completing sudoku puzzles, or playing chess or logic games. Dr. Soas also recommends not watching too much television because our brains go into a neutral state when we watch. (It's called the "boob tube" for a reason!)

If you don't have half an hour to spare, devise more creative methods of working out. When you go to the gym, tax your mind while you tone your body. Tune in to game shows like *Jeopardy* or *Wheel of Fortune* if you can. At the salon, flex your brain while you get a trim. Instead of staring blankly at the mirror, listen to a podcast or try learning a new language using your MP3 player. During your commute, play alphabet memory games like Betty White and Rue McClanahan did on the set of *The Golden Girls*. (Fruit:

apples, bananas, cantaloupes, dates; car models: Audi, BMW, Cadillac, DeLorean, and so on.) The key is to find new ways to make yourself think each day for twenty to thirty minutes.

Also remember that it's never to late to start. Contrary to popular belief, we don't lose huge numbers of brains cells as we age. "There isn't much difference between a 25-year-old brain and a 75-year-old brain," according to Monte Buchsbaum, director of the Neuroscience PET Laboratory at the Mount Sinai School of Medicine. In other words, mental decline isn't inevitable.

Disease usually causes severe mental decline, whereas most age-related losses in memory result from mental inactivity. Performing mindful tasks—such as reading mysteries or assembling jigsaw puzzles—can help ward off Alzheimer's disease and dementia. In other words, when it comes to your brain and its power, use it or lose it!

REMEMBER . . .

* Eat foods rich in omega-3 and omega-6 fatty acids.
* Your brain needs energy in the form of glucose from complex carbohydrates.
* Protein for breakfast kick-starts your brain into action.
* Avoid excessive amounts of caffeine, and drink plenty of water.
* Eat more frequent but smaller meals for better brain health.
* Exercise helps strengthen the brain.
* Brain exercise can delay the onset of diseases such as Alzheimer's and dementia.
* You're never too old to learn something new.
* Spend twenty to thirty minutes each day exercising your brain by reading, doing crosswords, completing sudoku puzzles, or playing chess or logic games.
* Watching too much television puts your brain into neutral gear.

PART TWO

GOVERN YOUR EMOTIONS, VANQUISH YOUR FEARS

✳

{12}

The Power of Emotion

W E humans are an emotional bunch, and our quest for success is an emotion-filled journey. But what are emotions, and how do they drive us toward our goals?

In technical terms, an emotion is a person's complex psychological and physiological response to a single stimulus or set of stimuli. The brain and body collude to create a state of readiness in response to a particular situation—fight or flight, for example. In humans, emotions fundamentally involve physiological changes, expressive behaviors, and conscious experience. Robert Plutchik, who was professor emeritus at the Albert Einstein College of Medicine, describes emotions as genetically based, unlearned behavioral adaptations that have value to their owner. They are patterned reactions as opposed to disorganized events. In simpler terms, think of emotions as the building blocks of feelings.

Emotions manifest themselves through different physiological changes to the body, the most obvious of which are facial expressions. The body also can respond with sweaty palms, raised hairs on the back of the neck, and alterations in heart rate and breathing. Ultimately emotions, as a subset of our minds, enhance our chances of survival and reproduction. According to Steven Pinker, Harvard University professor and author of *How the Mind Works:*

> the ultimate function of the mind is survival and reproduction in the environment in which the mind evolved—that is, the environment of the hunting and gathering tribes in which we have spent more than 99 percent of our evolutionary history, before the recent invention of agriculture and civilizations only 10,000 years ago.

But before we can have a physiological response to a set of stimuli, we must make a psychological evaluation of a situation. Before the hairs on the backs of our necks stand on end, our brains have to send the impulses for that to happen. For our brains to send out such signals they first must receive the appropriate stimuli to which they need to respond. These stimuli come from one or more of the five senses. So the process of emotional reaction goes like this:

1. Incoming physiological stimulus
2. Near-instantaneous psychological processing
3. Physiological reaction
Body first, then brain, then back to body.

Our brains process and decode incoming physiological stimuli extremely quickly. They must decide what to do in response: chase it to eat it, chase it to mate with it, or run away before being eaten. Based on the outcome of the calculation, the brain decides which parts of the body need the blood, oxygen, and nutrients to address the situation. Our brains take even the most complex physiological stimuli—such as watching an action scene in a 3-D movie on the latest high-definition, surround-sound-enhanced LCD television—and reduce it to fight, flight, or fornicate.

We're all capable of experiencing varying combinations of emotions. These include the six universal, or primary, emotions: happiness, fear, sadness, surprise, disgust, and anger. (There's debate about the exact number, depending on which group of scientists you believe.) These raw states of mind govern everything we do. Within them lie some 135 variants or blends of emotions. For example, within happiness, you'll find pride and confidence toward the anger end of the spectrum. Then, more centrally, come joy and pleasure. Toward the fear end of happiness lie such emotions as contentment and relaxation.

Emotional Building Blocks

Emotions consist of three different psychological components. First is enjoyment or pleasure. There's plenty of enjoyment in happiness but hardly any in anger, for example. Second, there's the component of excitement or stimulation. Emotions caused by confrontation contain a high percentage of excitement, whereas those of acquiescence contain only a minimal amount. Third is intimidation or a lack of control over a situation. Obedience and shyness both contain high levels of intimidation, while greed and impulsivity contain much smaller amounts.

These three components of emotion—enjoyment, excitement, intimidation—represent the mental switches that the brain uses to prepare the body for an instinctive course of action. Imagine that these components are the ingredients for making wine, grapes, sugar, and yeast. How much of each is used in proportion to the other two determines the taste of the wine. Too much of one ingredient, and the wine may turn out bitter. Get it wrong another way, and the mixture may explode from the barrel. Get the balance right, though, and you'll have a tasty and enjoyable beverage.

Emotions represent enormously powerful keys to our personal achievements both in terms of how we think of success and what our emotional reaction is once we begin to succeed. Once you understand what a powerful new set of levers you have, you'll understand more about how negative emotions are holding you back. Also keep in mind that emotion stimulates the mind three thousand times faster than rational thought, and the power ratio of emotion to reason is 24:1.

Emotions influence every decision we make every day of our lives. When you feel an emotion, you show it. When you feel sad, it shows in your posture as well as in your attitude. If you ask people to reflect on a sad event in their lives, you'll see a significant change in their physiology. Their shoulders will drop, the head slumps, and their breathing will become shallower. When someone thinks about sadness, his or her body exhibits the same emotion. Alternatively, if you asked the same person to adopt a proud, confident, upright posture and feel sad while doing so, he or she couldn't. Why? Because physiology and psychology work hand-in-hand with our emotions.

Emotions alter our mental and physical states, influence and drive our decision-making processes, and govern why we react and behave as we do. They can change our heart rates and level of physical activity. They can induce greater awareness of our environments and even cause excretions of bodily fluids—tears, for example. But how can you use this new-found knowledge to your advantage in your quest for greater personal achievement? Before you can make any decisions, you need to understand that your subconscious already has decided what's best for you from the perspective of fight, flight, or fornicate. Knowing that, you can harness those emotional urges to your advantage.

Exercise
Reprogram Your Emotions

Let's say that your success involves being a good stage performer. To get your brain onboard, you'll need to associate the thought of performing with the positive emotional aspects that accompany a solid performance. Along the same lines, you'll need to dispel any negative emotions such as fear.

To reprogram your emotional reactions, try some intense daydreaming. Good, impactful, emotional daydreaming involves powerful mental representations combined with an awareness of your own physiology and overall emotional response.

Imagine yourself performing as you want to be. Pay particular attention to the positive reaction of the crowd. As you build a mental representation of your onstage performance, pay attention to your emotions, both your mental and physiological changes. Keep playing your mental movie and focus on how it impacts you, noting only the positives. Doing this rewires your brain to believe emotionally that performing on stage is a good thing.

Continue this thought process as you consider how you feel after each great performance. Are you surrounded by people wanting your autograph? If so, you've strengthened your place in a particular social circle, which, from an evolutionary perspective, is good for survival. If members of the opposite sex are swooning over you as you leave the stage, then you've told your brain that performing on stage is good for its goal of procreation. Once you make learning to play music or recite lines more emotional, a larger part of your brain will commit to mastering the task.

While you fantasize about your success, you're reinforcing your brain's reasons for wanting to succeed. You're dealing with your brain on and with its own terms. A word of warning, though: This emotional rewiring is most effective when you prevent your rational mind from interrupting. Dream to your heart's content, and ignore practical realities as you do.

REMEMBER . . .

* Indulging in positive, impactful, emotional daydreaming will help reinforce your brain's desire to succeed.
* Making an event emotional will help your brain to make it happen.
* Define success in terms of fight, flight, or fornicate to maximize your brain's performance.

Social Emotions

THE parts of the brain that comprise the limbic system trigger and manage the universal primary human emotions. That's the system that helps us arrive at those familiar fight-or-flight, kill-or-be-killed, and procreation decisions. Almost all of the evidence suggests that these emotions lie beyond our conscious control. They occur automatically, having been hardwired into our brains since before we were human. So for any aspect of success you aim to achieve, you have to define it in those terms. Once you recognize the benefits from these basic survival goals, you can make your limbic system work with you to deliver what you want.

In addition to the six basic human emotions is another category, social emotions. Developed more recently, these include pride, respect, admiration, generosity, contempt, shame, and guilt. Unlike the universal emotions that govern survival, social emotions relate to how we perceive and are perceived by others. These more cultural emotions can serve as powerful allies when it comes to developing various kinds of relationships. Aligning your potential success by means of fight, flight, or fornicate is a good start, but associating success with social and cultural acceptability is just as powerful.

Social emotions aren't as old as their universal counterparts; they've been around only since our evolutionary ancestors decided to operate in groups. Estimates vary as to how old social emotions are. Some claim they began two hundred million years ago, when mammals first became warm blooded. Other estimates calculate it more recently, when anatomically modern humans originated in Africa about two hundred thousand years ago and then reached full behavioral modernity around fifty thousand years ago. Regardless of who's right about the timing, social emotions exist to ensure that we aren't excluded from our social circles or society as a whole.

Much of the success we want derives from how that success will improve others' perceptions of us. If you crave a designer watch, much of that desire comes from how you believe others will regard you when they see you wearing it. To train your brain to reach toward ownership, you first must approach the topic in the right motivational tone. Think through what you desire in terms of success, and then identify how that will make you feel. Ask yourself what the emotional consequences of attaining a particular success will be. Identify those consequences, and then confirm mentally that they're the resulting payoff. Doing so will help you decide whether that emotional state benefits your overall social well-being.

As we'll see in future chapters, discussions of emotions sometimes present two different problems. First are the semantics of language. Different cultures have different explanations of emotions, as do different people within the same culture. Also, evidence indicates that an emotion is an instantaneous response to the stimuli a person is facing, so they can be fleeting at best.

The good news is that while the limbic system—an older part of the brain—manages social emotions, the more modern neocortex—the reasoning part of our gray matter—also influences them. We humans have a more developed mental capacity than most other animals, which means we can examine our own emotional responses and cognitively assess situations so we can come to reasoned courses of action.

REMEMBER . . .

* To decide whether a success goal benefits your social well-being, identify how it will make you feel and what the emotional consequences are.
* When success comes, confirm that you feel as you expected.

{14}

Pleasure vs. Pain

EVOLUTION has hardwired a large proportion of our brains' activity, and the more modern parts, such as the neocortex, are woefully inadequate for the tasks we expect them to handle. As a result, our subconscious mind manages much of our decision-making processes. As much as 95 percent of decision-making happens below the level of conscious awareness—it's entirely automatic.

As we've seen, the basic survival responses of fight, flight, or fornicate drive our subconscious to its decisions, but other criteria acutely influence our subconscious as well. Our brains screen every decision we make for whether it will result in pleasure or pain. But our brains don't avoid pain and seek pleasure equally. Whenever we have a decision to make, our instincts go to the pleasure or pain filter, but they nearly always do more to avoid pain than to attain pleasure.

With this little bit of knowledge, you can review your decision-making more objectively. But to do so, you have to be aware of the decisions you're making—and remember that 95 percent of the time, they're subconscious. But making even 5 percent of your decisions more wisely will push you significantly toward your goals. Now you can look at a goal and identify whether it's likely to bring you pleasure or help you avoid pain, either mental or physical. If you can't decide, you probably haven't thought through the goal clearly enough.

When you do review your goals from a pleasure-or-pain perspective, pay particular attention to possible alternative courses of action that your brain may take en route. For example, you may have identified a goal that you think will bring you great pleasure. But your brain may have already decided that attaining that goal may bring you pain. In such a case, avoiding pain wins, and your brain will take the appropriate action.

Take this example: You want to run a marathon. Doing so will feed your self-esteem, deliver a number of emotional benefits as well as some physical ones, and allow you to spend hours pounding the streets with other like-minded individuals. Those are all pleasurable things. But training is going to be long, tiring, and probably painful. You might suffer some injuries along the way. Instinctively your brain will try to avoid the pain inherent in the goal by guiding you toward a self-defeating course of action, like cutting back on training. You have to recognize this form of self-protection and reframe the marathon so that the pleasure aspects at the end clearly outweigh the pain of getting there. Once you've done that, your brain is more likely to favor spending all those hours hitting the pavement.

Even when it comes to a relatively trivial decision, such as asking that girl or guy in the nightclub to dance, your instincts will screen your possible courses of actions. Your subconscious brain weighs the odds of a pleasurable romantic encounter (fornicate) against the risk of the social pain of being rejected or even the physical pain of a slap in the face. If you let your brain get away with it, you may decide that the safety of the periphery is preferable to the unlikely event of a kiss during the slow dance.

To avoid being sabotaged by your "pain brain," you need to reprogram your mind to look for the pleasurable positives rather than allowing it to dwell on the potential for pain. Obviously, pain-avoiding instinct sometimes proves right. You only have to watch home-video compilations to know that! Yet despite those participants' instincts—and occasionally family, friends, and even themselves—screaming "Don't do it!" emotions often win out, particularly when brain-altering chemicals such as alcohol help sway their decisions.

But by understanding how influential your brain's pleasure-or-pain filters are, you can better explain your previous decisions as well as develop a more refined way of deciding on successful courses of action in the future. Rewiring your goals to emphasize pleasure over pain can help drive you toward your ultimate success and help you use your emotional brain to make sure that those dreams of success have the greatest likelihood of becoming reality.

It's all well and good to map out your goals in chronological order, but if you really want them to materialize, you have to make them emotional. Emotions quickly, easily, and efficiently turn a desire for success into something that your brain craves, something it wants to help you attain.

When we set goals, we have emotional reasons for doing so, but when it comes to achieving them, we often rely on a more reasoned and intellectual process of getting there. That's where convincing your brain to stay on track proves useful. You can do that by providing the goal and the process of reaching it with emotional triggers and anchors.

Let's recap what we mean when we say "make it emotional." For your brain to understand how important a particular goal is, four things need to happen. First, you must feel some degree of pleasure or enjoyment both in your head and in your body. Second, the goal should convey a degree of stimulation or excitement, again both mentally and physically. Third, when thinking about a particular aspect of achievement, you should have a feeling of intimidation, which may be small but present nonetheless. Fourth, you should be able to generate a detailed mental representation of the achievement of the goal—you should "see" the results in your mind.

How do you add emotion to a goal? Let's start by asking why that goal is so important. What will the particular goal provide for you? Remember that your brain needs to know the emotional benefits, not just the physical ones, if it's going to help lead you to success.

Exercise
Make Your Goals Emotional

Write down on a piece of paper what you want to accomplish. Below that, list the reasons for wanting it. Then write down why that goal is important to you and what your emotional reasons are for wanting to attain it. Also note why you must attain that goal and what will happen if you don't.

Look at what you've written, and annotate every point you've made with an emotional reason. Make sure those reasons are brutally honest, the most emotionally meaningful, emotionally charged that they can be.

Now rank the points in order of their importance to you.

This is your blueprint for making a goal more emotional. If it's somewhat messy, no problem. That means you were emotionally involved while creating it rather than focusing on the neatness of what you were doing.

Read the list aloud, and sense how your mind and body feel about it. As you develop this skill, you'll be able to feel changes in your levels of pleasure, excitement, and intimidation. At the same time, you'll be creating even more detailed mental representations of each of your goals.

Remember, to train your brain to want it, a goal shouldn't be about the outcome as much as about what the outcome will mean to you emotionally. To illustrate the process, consider this example: A man wants to quit smoking cigarettes. Why? He wants to quit so he can live a healthier life. Again, why? Perhaps he wants to see his own children grow up and have families of their own. OK, that's good, but why? Perhaps his grandparents were there for him, so he wants to be there for his grandchildren. Wanting to help his grandchildren is much more emotional than just wanting to quit smoking because it's the healthier option. That's the emotional leverage that tells the brain that the goal must happen, that stopping smoking is mandatory.

The more emotional you get about your goals, the greater the likelihood that they'll become reality. Don't think so much about the process of setting goals and all the rational steps you need to make to achieve them. Tag them with emotion, instead, and know that your brain is working on attaining more personal achievement for you.

REMEMBER . . .

* You are wired to avoid pain more than to attain pleasure.
* Reframe your goals to avoid pain more than to achieve pleasure.
* Make your goals emotional to train your brain to want them.
* Imagine what the outcome of reaching a goal will mean to you emotionally.

I Am, Therefore I Think

A LMOST four hundred years ago, French philosopher René Descartes summed it up quite nicely: Cogito ergo sum. I think, therefore I am. Antonio Damasio, professor of neuroscience at the University of Southern California, has been thinking about it for years, and he finally decided that Descartes had it backward. It should be I am, therefore I think.

Whichever maxim has it right, both indicate that the human body influences thought and that thought influences the body. In other words, psychology and physiology work in tandem to create who we are and how we function. To experience an emotion that includes a feeling of achievement or accomplishment, you have to engage both your brain and your body. So let's look at how you can use the combination of brain and body to increase your personal achievement.

The conduit that links brain and body once again has to do with emotion. Emotion and feelings combine psychological and physiological states. Tony Robbins, author of *Unlimited Power* and a number of other success-oriented books, states that a person doesn't get depressed, he *does* depression. Ask someone to think that he's depressed, and observe his physiology: His head will drop, his shoulders will droop, the corners of his mouth will turn down, and his breathing will slow. That's because body and brain work together to generate an overall state of depression.

Conversely, when we think of excitement or surprise, our eyelids, heads, and shoulders all rise. Our mouths open, and we breathe deeply. This influx of air, combined with an increased heart rate, reflects our body's preparation for a form of action (fight, flight, or fornicate). Fascinatingly, when we're physiologically excited we *can't* feel depressed, and when we're depressed we can't feel physiologically excited. That means that to experience a feeling of achievement, we have to engage both brain and body on the same track.

Let's return to the Damasio quote. He said in his book *Descartes' Error* that Descartes had it wrong when he created "the abysmal separation between body and mind, between the sizable, dimensioned, mechanically operated, infinitely divisible body stuff, on the one hand, and the unsizable, undimensioned, un-push-pullable, non-divisible mind stuff." Mind and body can't be separated because they work together as a team. Many now believe that this combined brain-and-body functionality is what drives the human mind, a suggestion supported by Steven Pinker in *How the Mind Works*. But Damasio and Pinker were hardly the first to arrive at this conclusion. Charles Darwin similarly summed up the situation: "The same state of mind is expressed throughout the world with remarkable uniformity; and this fact is in itself interesting as evidence of the close similarity in bodily structure and mental disposition of all races of mankind." Darwin also believed that the mind comes from the combined activity of the brain and body.

Once we accept this premise—that emotions, feelings, moods, and states of mind come from the combined efforts of the two—we can apply that concept to our achievement strategies. First, we need to review what it feels like to be successful. To do that, we need to identify what success means to us personally. Remember, we're more emotional than we are rational, so the best way to identify success is in emotional terms (contentment or pride, for example).

Let's take pride and start at the beginning. To experience how pride feels, you need to identify a proud person and then model your behavior after his or hers while thinking about how and what that person is thinking. Find videos of proud people, and look closely at their physiology and posture. How they carry themselves says it all.

As another example, search out professionals, such as actors or sports stars, and see how they react when they receive an award. They're receiving public recognition for their achievements, and at the same time they're exposing their emotions to other people. Now internalize their behavior and feelings of being a winner. What you're doing here is developing your own combined physiological and psychological strategy for achieving success. You're learning from others' reactions how to be more successful.

To carry that example forward, every time you want to feel that winning feeling, play out your strategy by physiologically modeling the success of others. Think of who they are and how they handled the accolades.

But most of us already know how pride and contentment feel, so let's take the next step and look a little more closely at how success feels.

REMEMBER . . .

* Psychology and physiology work together to create who we are and how we function.
* To achieve success, you have to engage both your brain and your body.
* Emotion is what ties them together.

{16}

The "Feel" of Success

As we've seen, different forms of happiness exist, and they cover contentment, ecstasy, joy, pride, and trust among other variations. The single constant factor in whatever type of happiness you seek or feel will always be enjoyment or pleasure—far in excess of any excitement or intimidation that you also might feel. For the purposes of maximizing your success for achieving happiness wherever and however you want, let's take a look at how to increase enjoyment when contemplating a situation.

You must communicate whatever you desire in life to the appropriate parts of your brain in a way that is congruent with your emotions, both primary and social. Whenever you think of what you want, pay close attention to how that desire manifests itself within you. What's your mental sense of the desire, and how does your physiology change when you consider what will happen once you achieve it?

Take, for example, the goal of becoming more financially secure. It's not the money itself that will give you financial security. Money is worthless until you spend it. But money does act as a conduit to what you want to achieve. Let's say you want to be better able to care for and protect those you love. Worrying about how you're going to do that isn't as constructive a message to your brain as communicating to it the sense of achievement from knowing your loved ones are secure. Focus on how proud you'll feel knowing that you have provided the money needed to insure your loved ones have food, shelter, and safety. Then your brain will help you develop a clear and congruent strategy for delivering on your desire.

When you pay attention to the happiness you want, focus on what sort of happiness appeals to you most. Perhaps happiness is the pride you feel at the respect from your social circle. Alternatively, it could involve being loved by others in a variety of ways. It might be that you want to simplify your life so you can view the world around you more calmly. In each of these situations, your desired emotional state will be different, so begin by coming to grips with how that emotional state manifests itself. The clearer the emotional message, the more straightforwardly your brain can help you deliver on your desire.

But not everyone knows what success feels like. Some people are so negatively wired that they have trouble experiencing or even imagining happiness. If you're not sure how happy you are, here's a short test that will help you tell.

<hr/>

Exercise

Is Your Body Happy?

What's the opposite of success for you? What are you trying to avoid on your quest for achievement? These are your fears. Take a moment to think about them. How do those negative thoughts make you feel? Now identify what you've already done—in the past month, in the past year— to counteract or prevent them. Next, envision the opposites of those negative thoughts. These are your hopes and goals.

As you focus on your achievements, don't overlook your physiology. If your shoulders are slouched, pull them back and push out your chest. If your breathing is shallow and weak, take several long deep breaths. Observing yourself in a mirror can help weed out negative body language. Crossed arms? Uncross them. Bowed head? Chin up. Tense expression? Relax those tight muscles. What's your facial expression telling you? If you have a worried look about you, correct it. If fear shows in your eyes, relax your face and smile.

<hr/>

That's a relatively easy fix—your physiology. Now let's move on to something a bit tougher, your mental state. Once you've altered your physiology, you may begin to sense that you're also in a better place emotionally. Here are a couple of additional suggestions that scientists have identified as being effective emotional management tools.

Think about the root cause of your negativity: fear, sadness, guilt, or what have you. Now think of the upside, the pleasure that eliminating that negativity will bring you. For example, if you fear asking someone on a date because you don't want to experience rejection (your brain protecting you from pain again), imagine your emotional state as you walk arm-in-arm with that gorgeous person in some idyllic setting, such as on an isolated mountain trail or sandy beach. What you're doing here is making the upside of a goal feel more emotionally enjoyable than the downside feels negative.

Perhaps you're nervous about asking for a raise. If so, you need to reframe your focus. Instead of thinking of your boss scowling at you from across the desk, think about being able to afford more of the things you desire and the status you'll receive from climbing the social ladder.

In both of these examples, you're communicating with your brain in the ways it has evolved to function. By doing so, you're preventing your mind from misunderstanding what you want from life and increasing the likelihood that it will prompt you to what you want.

A key tool we've been using is envisioning how success feels—but if you don't know what personal achievement feels like, you can't advise your brain on what you want. If you don't know what emotional target you're aiming for, you won't ever hit it. But personal achievement means different things to different people. What you want to achieve is likely different than what my goals are. In fact, many different types of success exist, but they all fall into two different definitions that anyone determined to achieve them must embrace.

The first definition refers to what it is: the achievement of something desired, planned, or attempted. A second, more refined definition is the reaching or surpassing of a goal or target set. When we look at success that way, we face an interesting question. Do we have a clearly defined finish line at which we can say with certainty that we've achieved our particular success? If you do, then great, but for many success represents a woolly dream, way out there in the future. It's something like "Every day, in every way, I'm getting better and better," as French psychologist Émile Coué put it. It's an intangible—you'll know it when you achieve it.

There's nothing wrong with that philosophy, but it is, for the most part, ineffective. Take the "getting better and better" part. That might work just fine if you're in bed with the flu, but if you're seeking more tangible forms of success, you're going to need a more embraceable definition.

To use an analogy, an airline pilot enters destination coordinates into the navigational computer, and then the autopilot program guides the plane to that location. Without correct and specific coordinates, the autopilot system would be useless. It's no good leaving JFK for Heathrow and punching in, "To the west and up a bit." Your brain functions like an advanced autopilot device, its navigational effectiveness relating directly to the quality of the coordinates you enter. On your quest for success, you need to know not only what success is so you know precisely when you've reached it but also what it feels like so that you can give your brain the right mental coordinates for the destination.

Consider the example we used earlier in the book. If you've always wanted a shiny new Mercedes Benz and one day it's sitting in your driveway, you can check off that item as a success achieved. Perhaps you've long dreamed of waking up in the arms of someone you love and one day you finally do. That's another achievement off the list. To be able to achieve success, you first must recognize what it means to you and have a specific point and clear definition of what achieving it represents. That way you're telling your brain what you want and then rewarding it by recognizing when it delivers on your goals.

A moment ago, we looked at the two different definitions of success, but how did you feel when you found the man or woman of your dreams? Knowing that you need to feel that sense of achievement is important, but defining the actual feeling itself is even more so. That's because our brains constantly monitor and refine how we feel from moment to moment. What provided feelings of euphoria three weeks ago might trigger only a minor sense of elation today. Remember, your emotional brain has been evolving for far longer than your more modern, reasoning brain, so it responds more effectively to emotion than to reason. That's why you need to make your personal achievement emotional. Don't just think, *OK, I got the car, just like I always wanted.* Instead, *feel* your reaction. *Whoo-eee! I did it! I can't believe it—my very own Mercedes Benz! Just like I always dreamed of owning!*

Just as you create clear representations in your mind as to what success is and what it looks like, so should you spend time building a representation of how success feels. Dissect that feeling into how your mind perceives it and how your body reacts to it physiologically. Pay attention to heart rate, breathing, poise, and posture while you think about achieving your success. By doing so, you're communicating to your brain in a language it already knows.

REMEMBER . . .

* Focus on what kind of happiness appeals to you most.
* Pinpoint your success milestones.
* Recognize success and how it feels.
* Identify the root causes of any negativity you feel.
* Make the positives of a goal feel more emotionally enjoyable than the downsides feel negative.
* Congratulate yourself emotionally when you achieve your goals.

{17}

Listen to Yourself

WHEN you remember a past event or dream of something in the future, how do you do it? That is, what's the tone of your inner voice? Not sure? Then let's see what you notice about the following two paragraphs.

> When I arrived, I was annoyed to see a long line, and my frustrations grew as I waited for more than half an hour. When I finally got inside, I was bumped and jostled by a never-ending stream of people trying to get the best vantage point. When I found a decent spot myself, although bedraggled and drained of energy, I prepared for the main event. *At last*, I thought to myself.

> The first thing I noticed when I got there were loads of people, just like me, who couldn't wait to get in. We all laughed and joked as we approached the admission gate. My heart was racing, and the excitement grew as we all filed through the turnstile. *This is it!* I thought. I was in! And it wasn't disappointing. The tidal wave of expectation carried me along until I found myself in a great spot to see the show. I thought I was going to burst with anticipation before it began!

Here we have two very different reactions to the exact same event. In the first example, the tone is negative, filled with frustration and negativity. In the second, the language is much more positive and upbeat.

When it comes to describing events past, present, or future, see if your cup is half full or half empty. Do you see the upside or the down? Here's an exercise that will qualify how you replay your memories. You can use it for past events as well as to create a mental representation of something in the future.

Exercise

What's in Your Cup?

Find a place where you can spend half an hour undisturbed—somewhere you can write in comfort. Think of a meaningful event from your past, a birthday perhaps or a family event. Write down, in as much detail as you can, a description of the event, good, bad, or otherwise. What lead up to it? How did you feel during? What was going on around you? What ultimately happened?

Don't worry about neat writing, proper spelling, writing style, or the like. Just keep writing. The goal here is to get down in words how you think, so you can identify the tone of your inner voice.

When describing the event, add as much detail as you can, and make sure to reflect on how each detail made you feel. At the end, write down how you felt afterward and how you feel about the event now.

Once you've finished, you have a written version of how you talk to yourself, including style of words, degree of positivity, and even your representational system.

Wait—your what?

When you remember a feeling or an event, you use your representational system, which is how you use your five senses to recreate past events from memory. When people say they talk to themselves, they are referring to an auditory representational system, even when no words are spoken.

When a person creates an image in her head when thinking or dreaming, she's using a visual representational system. When someone describes physical movement, that's a kinesthetic representational system.

Now let's go back to what you wrote. Underline or highlight every negative word in red: *angry, annoyed, bad, can't, fail, frustrating, no, poor, problem, sad,* and so on. At the end, count how many you have and note the total.

Next, go back and repeat the exercise, highlighting or underlining each positive or upbeat word in green. Look for: *active, certain, content, easy, fun, happy, lovely, proud, rich,* and *yes.* Count up the number of positive words and note the total.

Compare the two numbers. The higher the positive number, the more positive your internal dialogue, and vice versa. Once you've identified how positively or negatively you tend to think, you can make yourself more optimistic. Read your text again, and look for negative words that you use more than once. Cross out the negative phrases and replace them with more optimistic or positive words.

Now go through your entire text and replace every negative word, phrase, and sentence with a positive alternative. Read the whole thing back to yourself. How does it sound? Better? Stronger? More positive?

You now have a tool to make yourself more optimistic wherever and whenever you want. You can extend the process to letters and e-mails anytime you communicate with others in writing. The resulting "new you" will impress people by how upbeat a person you are! Who knew?

REMEMBER . . .

* Using positive, upbeat language will help you feel better.
* Reframe negative words by replacing them with positive ones.

⟨18⟩

Emotion and Memory

WE don't always like to admit it, but emotion constitutes an integral part of memory, both in terms of what we remember and what we forget. Earlier we examined how our brains remember things. Now let's apply that knowledge and discover how emotion holds the key to so much of what each of us remembers.

Memories are ingrained in our personalities. We remember what happened and how it made us feel, but more specifically we remember what happened *because* of how it made us feel. Think back to a memorable event. It might be an unexpected argument, a rock concert, a momentous birthday, or a memorable vacation. Recall the details, including, the sights, sounds, weather, and other information relating to the event. Doubtless your list of details exceeds a hundred different points of information. When it matters, we all have astonishing memories.

Now let's try another test. In alphabetical order, name the fifty states. For most of us, this is much harder—even though we've had years of formal education and probably memorized that information at some point when we were young. Memory itself isn't about whether we have good or bad memories but rather about how we file information into them in the first place. A key way to get information to stick in your memory is to tag it with emotion. Tie together even the most mundane points you're trying to remember with how you feel when you think of them, and they'll last much longer in your head.

Instead of reading and repeating the names of the fifty states by rote, imagine having a romantic meal with your partner. (Follow me on this, it works.) Lean forward and whisper, "Alabama," and she or he responds by gently caressing your hand and replying, "Alaska." Take hold of his or her hand and respond, "Arizona," and the response comes, "Arkansas." Yes, it's an extreme way to recall a mundane list, but you're tagging each state with a specific emotion and feeling, so you're filing each of those names into your memory. Because the list builds a romantic storyline, you'll sequence it correctly in the future.

As a general rule, we remember emotionally charged events better than emotionally neutral ones. The latest research suggests that it's the emotions aroused—not the personal significance of the event—that make those events easier to remember. In other words, we remember emotions and then associate them with events more accurately than recalling the events themselves.

From how success feels to mundane lists of places, now you have a surefire way to improve your ability to recall information. Understanding what your brain is doing as it consigns points of reference to memory and, more importantly, how it indexes that filing system gives you a significant advantage in life.

Another aspect of what we remember or forget has to do with whether our memories are pleasant. Pleasant emotions fade more slowly than unpleasant ones. Which means we really do see the past through rose-colored glasses. Our brains do what they think is best for us so that we're always replaying the positive in our brains and minimizing the negative.

Occasionally, I run across someone who pays way too much attention to his or her previous failings—to the point of dwelling on them. If that's you, think again. Yes, we all want to learn from our mistakes, but the key is to learn, not to relive. From an emotional perspective, learning comes from reframing a memory so that you can pull something positive from it, however small, and focus on that as opposed to the negative aspects.

All of us can frame memories the way we want to perceive them. For example, someone I know got into money trouble some years ago. He fell behind on paying his taxes and had a yearlong struggle with the IRS before finally having to close his business. As a result, he developed a powerfully negative belief based on distorted memories packed with negative emotions and feelings that his lack of success came from the nasty government officials who had stifled his creativity and ruined everything. Never would he consider starting another business just to give it all away to the tax man!

We set about reframing his memory, for a start, by identifying others who had experienced the same setback. The more noteworthy include: Charles Goodyear (tires), Walt Disney (film), and Henry Ford (automobiles). We concluded that life does exist after bankruptcy, which doesn't have to mark you as a failure for the rest of your days. For many people, the challenge of rising again made them that much more determined to succeed in the future. Plus, on a more practical note, people who know how failure feels often vow to target success more stridently, which, of course, the brain finds far more palatable.

Once my acquaintance reframed his memory, it became apparent that the IRS had provided him an invaluable education about how to manage tax affairs in the future. He now is a successful businessman in charge of a number of global companies. How can I be so sure that the negative experience transformed him? Because that person is me.

When we need to remember information, we have to recall the emotions surrounding it. We also need those emotions to be good rather than bad, but sometimes we can't avoid negative emotions or the negative memories that they form. Let's take a look at how to undo some of their power over us.

We retain memories, bad or good, because the lines of communications between those particular synapses grow stronger and thicker the more we activate and access them. The first key to addressing bad memories is to recognize what comprises them.

Unfortunately, bad memories can become so powerful that they prevent us from doing what's best or even in some instances what's right. So we need a method for realigning our brains and those memories so we can follow our quest for success. We need to learn that what we remember doesn't have to have the same power to influence us now and that wiser courses of action exist.

The Memory Compactor

Here's a technique for disarming those negative feelings by identifying what triggers them and then dismantling them until they don't have the power to debilitate you anymore. It's not a magic means by which you can erase your worst memories, but it is both practical and user-friendly. The process has two parts: First we'll learn how to disable a bad memory so it's less painful to recall. Then we'll reframe an aspect of it to divert your focus away from the parts that may have become psychologically harmful. A word of warning, though: for some this won't be easy because first you have to call up the memory before you can pull it apart. But if you complete the exercise properly, the memory should never have the same damaging hold over you again.

Let's begin by calling that bad memory to mind. What causes you to think of this negative episode? What's the trigger? If you can identify it, then good—but if not, don't worry. We're going to deal with the entire memory anyway.

First, you need to disassociate yourself from the action of this mental representation. Instead of being a part of the scene and looking at the episode through your own eyes, step outside and look at it from a distance. Imagine that what you're viewing is playing on a screen in front of you and that in your hand you have a remote control. Play around with the remote control and notice how you can pause, rewind, and even stop the action on the screen altogether. Hit the mute button, and hear how the memory goes silent. You now have no negative noises left to interfere with the representation. Another aspect of the control you now have over this memory is that you can control how you feel watching it. If it used to make you cold, you can now feel warm.

Address all of your senses as you control how you feel while watching the memory on the screen. Focus not on how the original event made you feel but on how being able to manipulate it alters your perception of the event. The more you focus on operating the remote rather than the original event, the more you will disassociate yourself from that episode in your life.

Now let's have some fun with it. Press a button on the remote to change the color scheme to black and white. Press another button to weaken the signal, and see how the picture onscreen goes fuzzy and indistinct. If any sounds are still there, mute them again. If they're still coming through, press another button on the remote to change them into something idiotic. For example, change a fearsome voice into the noises that the Tasmanian Devil cartoon character makes or alter it so it sounds as though the person has just inhaled from a helium balloon. Spend some time on this, and enjoy yourself as you get creative with the special effects of this mental representation.

Now for the next part of the process. Redirect the signal even more until the image consists of nothing but snowy white noise, like disconnecting the satellite dish. Pull the power cord from the back of the screen so it goes black. Take a hammer to the screen and shatter it.

Finally, envision someone else—not you—taking the shattered screen and remote control and depositing it in a compactor. Watch as the compactor crushes it into tiny pieces. Hear the noise as it breaks and the casing splinters apart. Smell the acrid scent of plastic or glass as the screen is destroyed completely. Then see a tiny little lump fall out of the compactor, no bigger than a matchbox.

At the beginning of the exercise, I asked you to recall the trigger that brought this memory to mind. Think of that trigger again, but this time pair it with the mental image of the crushed screen and remote falling, small and ineffectual, from the compactor.

At first, this process may seem a bit odd, but all we're doing is dissociating you from the memory and then reframing it. Like everything in this book, the more you practice and the harder you try, the more effective the technique will become.

REMEMBER . . .

* The feeling of an event is easier to recall than the details.
* Tag emotions to information to better remember it.
* Learn from your mistakes rather than reliving them.
* Focus on the positive, and ignore the negative.
* Bad memories consist of smaller component parts. Disconnect the components, and the bad memories will fade.
* Dissociate yourself from and reframe bad memories to debilitate them.

{19}

Emotional Priming

OUR short-term memory can store information for no longer than eighteen seconds. Within that time, our brains are constantly analyzing, filtering, and discarding much of the incoming sensory stimuli they receive, and our mindsets and feelings are constantly adjusting in response to the latest stimuli. Our moods continually fluctuate from one moment to the next, depending on our environments.

But our moods and feelings aren't a set of on-off switches. They're more of a moving line on a graph, rising and falling in waves. We don't receive one pleasurable stimulus from one of our senses and then become instantly happy and joyful. A response depends more on how we were feeling at the moment the stimulus arrived in our brain. If we were sad, the new stimulus might only make us less sad instead of deliriously happy. Conversely, if we were already happy, the stimulus could make us happier but maybe not as much as if we were sad to start.

Left alone, the stimuli entering our brains linger in our conscious for no more than eighteen seconds, assuming of course that additional stimuli don't come along and capture more of the brain's attention. It's always a case of how we felt influencing how we'll feel. Which can be both good and bad for your personal achievement outlook.

On the positive side, you can prime yourself to feel a certain way when you know it's important to do so. For example, just before going into that big job interview, focus on how you'll feel the moment you are offered the position. As always, develop as clear a representation as you can, remembering to place yourself within the representation rather than watching yourself from the sidelines. When you're standing in a bar or club trying to find the courage to introduce yourself to somebody, concentrate on how it will feel to stand arm-in-arm with that person.

Another success technique involves having a trigger with you when you want it such as a picture of someone or something you crave. If you have a baby or a child whom you treasure, carry something small that he or she has made into important meetings or wherever you need to perform at your best.

Many people benefit from placing images of loved ones in their wallets and offices so that they see them while working or running errands. But what about taking that further? By all means, have that image in your pocketbook or car, but add to it by having the person record a motivational or good luck message. Then you can play it on your MP3 player whenever you need the extra inspiration. You can also create a mobile phone ringtone from a snippet of that message that will trigger the emotional response you want.

You can also use other ringtones with other people in your contacts list. Spend some time mentally associating a sound with the emotional state you want to have with a certain person. Then, whenever he or she calls you, you'll automatically enter the state of mind that you've programmed yourself to activate.

Why does this work? Music—including ringtones—can influence your brain. According to research by neurologists in Montreal, music can arouse feelings of euphoria and craving. The results of several studies indicate that intense pleasure in response to music can lead to the brain's release of dopamine, a brain chemical associated with pleasure and reward. Ringtones are a great way of harnessing that system.

———

We can prime ourselves in good ways and bad. Take an audit of sensory cues around you, and identify whether they are beneficial in your quest for success.

For example, habitually watching the news is neither conducive to positivity nor personal achievement, so avoid it and read the day's headlines and stories instead. As you conduct your sensory audit, look for any items that cause negative or limiting beliefs, including reminders of past failures that trigger negativity or gifts from an ex who hurt you. Pictures on the mantel, clothes in the closet, jewelry on a dresser, or even a car—check them all out. Wherever possible, remove those damaging mental triggers and replace them with something more positive. If you invest time and effort in conducting this exercise, you'll be amazed at how many items in your day-to-day life have a tangible influence on how you feel. Once you've addressed them, you can choose whether to discard them, hide them, or at least reframe their meaning to your brain.

How many times have we heard of people wanting to make a fresh start in life, yet when they do they're still unhappy? That's because their negative triggers still feature prominently in their lives. The key, then, isn't so much throwing away the past and starting over, as it is something more scientific. Look for the triggers that led to the limiting mindset in the first place. It's less expensive, less time consuming, and more effective to identify the aspects of the past that cause us mental harm and deal with those. Don't forget, there's a lot of good stuff in our pasts, too.

REMEMBER . . .

* Display an image of a loved one somewhere prominent to inspire you toward success.
* Even better, have that loved one record a motivational message that you can play whenever you need it.
* Use ringtones to put you in the right frame of mind before talking to someone on the phone.
* Carry out a sensory audit of what surrounds and impacts you.
* Minimize or remove any negative sensory reminders from your surroundings.

The Science of Confidence

THE one emotion that nearly all successful people appear to call on more than any other is confidence. But what is confidence? And if we don't have it, or don't have enough of it, how do we go about getting it?

According to the *Oxford English Dictionary*, confidence is "the feeling or belief that one can have faith in or rely on someone or something" and "the state of feeling certain about the truth of something." In other words, confidence is certainty that a hypothesis, prediction, or theory is correct or that our chosen course of action is or will be the best or most effective. Those of us determined to achieve more need to have self-confidence.

We saw earlier that every emotion consists of a certain percentage of enjoyment, excitement, and intimidation, the last of which can also be seen as loss of self-control. The main ingredient in confidence is enjoyment. The second is excitement, with intimidation bringing up the rear. For the mathematicians among us, the equation for confidence is enjoyment minus 15 percent excitement and minus 95 percent intimidation. In less numerical terms, when you think of a particular course of action you'd like to take, what do you feel most? If you're nervous, the leading component is either intimidation or excitement. To find out which, you need to look at your own emotional state and identify how and why you feel the way you do.

The reason for your intimidation may stem from a fear of failure and embarrassment in front of your peers, or maybe you fear physical pain. In either case, your brain is working to protect your own interests even at the cost of personal achievement. To counter those fears, you need to make the thought of taking a specific course of action more pleasurable. Identify and focus on the enjoyable aspects of the outcome from taking the action.

For instance, suppose you're considering meeting with your boss to ask for a promotion. Chances are that the thought of such a meeting fills you with nervousness or even fear. What you have to do to feel more confident is to refocus on the pleasurable results of the meeting such as more money and greater workplace respect. Think about the situation in terms of how you'd feel in your new role, as if you already had been promoted. Pay attention to how you perceive this situation mentally and how your body adapts physiologically. Spend a few minutes developing this mental representation of yourself in your new role. As you do, you're communicating to your brain how you want to feel and how the achievement will make you feel. As a result, you're naturally less able to pay attention to the excitement or intimidation you feel. The more you associate pleasure with a course of action, the less you'll feel hesitation, nervousness, or fear.

Here's an example. My young cousin was suffering from extreme nervousness about a number of tough high school exams coming up. Fear had so engulfed him that his entire quality of life and outlook for the future had become twisted and distorted. I had him refocus his mind on the adulation he'd receive from family and friends for performing well on the tests. After talking with me for a few minutes, he concentrated not on the fear of opening the exam booklet and drawing a blank but on the joys of scoring well. As he did so, his entire physiology—his body language and persona— changed before my eyes.

But one particular exam still made him nervous. For that, I gave him added ammunition. I promised him a small financial reward for each exam he passed. If he passed them all, I offered a disproportionately larger financial incentive. But there was a catch, and it was the catch that made his brain hone in on achieving the desired results. Whatever the outcome, he had to tell me either face-to-face or over the phone. That made him concentrate on how he'd feel when telling me how he did and reminding me that I owed him money. In this case, he passed every single exam and eventually enrolled at the college of his choice. Me? I was out a few dollars, but the loss was really a gain for us both.

Don't think of a lack of confidence as an insurmountable obstacle. It's not. If you reframe your emotions by focusing on the emotional benefits of taking a particular course of action, your confidence level will soar. Identify the upsides of making that public speech, asking that special someone out on a date, or scheduling a meeting with your boss. As you do, pay close attention to the pleasure that you will feel as a result, not on the negativity that might happen.

―――――

But sometimes confidence is less about action and more about how you view yourself.

None of us is perfect. We often focus on our imperfections, assuming that everyone else is concentrating on them as well. But most people are worrying far more about their own flaws or deficiencies than others'. So we can choose whether or not to focus on our own imperfections. We're not doomed to dwelling on our own shortcomings, real or imagined.

But some people think about themselves emotionally, rather than rationally, in a way that causes serious damage. They let their imperfections become all consuming. Their obsessive thoughts turn into distorted beliefs. The brain can make incredibly complex calculations and reach decisions in mere fractions of a second, but it also misjudges a lot. Often what others consider a positive trait, we ourselves might think of as a defect.

A woman's body makes an interesting example. Men's evolutionary preference for curvier women has a lot to do with shapely women containing more of those omega-3 fatty acids essential for proper brain development in children. Biologically men "know" something significant about women's bodies that women don't.

The average woman is more the average man's ideal than she realizes, and losing weight to meet the benchmarks set by the fashion world—a multi-billion dollar industry designed to sell clothes—probably won't make her more attractive. Men rate women as most attractive when they have a waist size around 60 to 70 percent of their hip size. The typical female American college student has a waist that's 75 percent of her hip size, while the average fashion model's is 46 percent.

Much of a woman's fat store coalesces on her hips, butt, and legs. During the last few months of pregnancy and after childbirth, her body starts breaking down this lower-body fat, making it available to the growing fetus or infant. Babies' brains grow fastest in the first two years of life—but only if they have an adequate supply of those omega-3 fatty acids. See the connection?

So the next time you compare yourself to others and feel self-conscious, stop your brooding by reframing the focus of your attention. Consider what they might be fretting about regarding themselves. By focusing on others, your brain pays less attention to you, and when you realize that other people aren't perfect either, you'll think more favorably about your own self-image.

Exercise
Build Your Confidence

Answer the questions below, and every time you need to feel more confident this exercise will give you the boost you need.

List five tasks that you have done well in the last twenty-four hours. It doesn't matter how minor they are, but they must be undertakings that you were responsible for doing right.

1. _____

2. _____

3. _____

4. _____

5. _____

Now list five tasks that you've done right that have resulted in something positive in your relationships with other people.

1. _____

2. _____

3. _____

4. _____

5. _____

Finally, list five life-changing tasks that you've done in your life that, in hindsight, were the right thing to do.

1. _____

2. _____

3. _____

4. _____

5. _____

Now read your lists aloud, and visualize each item as you say it. Any time you need more confidence in yourself, read the lists aloud again and visualize each item.

REMEMBER . . .

* Confidence equals enjoyment minus intimidation, minus some excitement.
* Lacking confidence can be overcome.
* Focus on the emotional benefits of taking a particular course of action to overcome hesitation, nervousness, or fear.
* If you're feeling self-conscious, take your brain's focus off yourself by imagining what those around you might not like about themselves.

{21}

Shortcuts to Failure

W E'RE starting to form a pretty good idea of what we need to do to transform a burning desire into success, including overcoming a lack of confidence. Here and in the next several chapters, let's look at some other stumbling blocks we must overcome.

One of the worst is apathy—lack of interest, enthusiasm, or concern. It's that sullen, angst-ridden attitude that says, "I don't care." Sitting side-by-side with apathy are excuses. They're an attempt to lessen blame in order to defend or justify some course of action, including no action at all. If you have a lack of interest in success, then your brain will probably attempt to justify your inaction.

If you have any lingering doubts about how much you want to achieve success, you need to look more closely at why you lack the ability to really want things. Without that burning desire, you'll find it difficult to generate sufficient enthusiasm for the goals you set yourself.

We've all experienced the pleasure of some kind of achievement in life. To increase your desire toward a particular goal, you have to call up that sense of achievement, which will help you create a more intense desire for success. Unfortunately, what lies behind most of the apathy and excuses in our lives is little more than our past failures. Our brains make excuses by focusing on negative outcomes from the past, taking the negativity of previous shortcomings or failures and magnifying our emotional response. As a result, our brains—always avoiding pain, remember—are less likely to try again. To battle that tendency, we need to understand that failure doesn't exist, only a different outcome than the one we wanted.

No one gets everything right every time. Everyone makes mistakes, and all of us have experienced how it feels to fall short. When that happens, our brains often deploy apathy and excuses as defense mechanisms. They reduce the importance of attaining a goal (sour grapes), thereby making us less determined to try to achieve it again. Then they shift blame to anyone other than ourselves. We may decide that not reaching the goal was society's fault, a parent's, a coworker's, or a friend's.

Imagine, for example, that you want to lose a few pounds. After several dieting attempts, you lose barely any weight, which leaves you frustrated. Rather than admit that some component of the attempt was ineffectual, your brain tries to save face. It floods you with memories of past failed diet attempts. If you failed in the past, why bother trying again?

A better train of thought is to concentrate on the weight that you did lose or the muscle you gained, however small, and to recognize that as a positive outcome in line with your overall goal. That way you can communicate a different message to your brain. *OK, I lost some weight or converted some fat into muscle; now I have to figure out how to do more of that.* It's not so much a revolutionary new strategy for losing weight, as it is a refinement of an already proven strategy. Don't let your brain trick you into thinking that because you didn't lose all the weight you wanted the entire diet was a flop. But that's exactly what your brain will do.

It's important, then, to fortify your successes—plural—by taking smaller steps instead of trying to reach your success goal in one giant, Olympic leap. In the case of dieting, that may mean eating less fattening foods or eating smaller portions rather than eliminating certain food groups altogether or cutting your caloric intake in half. Choose an easy, realistically achievable first step that will offer the least resistance and therefore the greatest chance for success. That may mean replacing the sugar in your tea or coffee with a low- or no-calorie sweetener or switching from butter to a low-fat spread. The aim is to persuade your brain to stop prevaricating and do something— anything!—constructive. That will start a chain of events that will convince your brain more effectively to forget those previous unsuccessful attempts.

Your brain works overtime to protect you from pain, often by reflecting on past failures and avoiding any action that may result in the same outcome. You can short circuit your brain's hardwired negativity by disarming its apathy and excuses. Whenever you catch yourself sidestepping a course of action toward a goal and you sense that past failings are to blame, make your resolve even harder to take that first step. Make it a little one, and not a quantum leap, and you'll amplify your chances of success.

REMEMBER . . .

* Don't allow apathy or excuses to drag you down.
* There's no such thing as failure; it's just a different outcome.
* Focus on successes, however small, more than dwelling on shortcomings.
* Break difficult goals into smaller, manageable steps to help you achieve them.

{22}

The Past Is History,
the Future Starts Now

A S we continue looking into obstacles that prevent you from reaching your success goals, let's look more closely at how your past influences your future. But instead of discussing your past externally as previous experiences, let's look at how you think about it. Some people are more positive than others and express themselves in a more optimistic voice, while other people's perceptions of the past and future vary. Some dwell in the past, for good or bad, while others focus on the future.

The more of a past you have, the more you're inclined to refer to it. As an example, children talk more about what will happen in the future when they grow up, whereas retirees more often mention their earlier lives. Your preference can have a significant impact on your outlook toward your own achievement. For example, do you think of success in terms of the benefits it will provide you in the future or how it might alleviate problems you've already experienced?

Historians aside, many people who live in the past don't realize that they do. They inadvertently allow past achievements or shortfalls to shape their future. That's not necessarily bad; after all, who doesn't want to learn from past mistakes? But many past-dwellers tend to overvalue past events because they're giving insufficient thought to the future and its opportunities. People who focus too much on the past constantly mention it as though it matters just as much now as it did then. A simple technique to avoid this linguistic behavior is to first note when you're speaking in the past tense and then to calculate the number of years between the event you're describing and now. It might not feel like twenty or even fifty years ago, but calculating the number will help draw your attention to the habit. Then concentrate on what you want to have achieved or what you want to have happened the same number of years into the future. Over time, you'll find that you're replacing those past references with more positive future ones.

Of course, not everyone lives in the past. Those who live in the present and for the moment often have scant regard for history or the future. Again, it's not wrong to make the most of your current situation, but if that's all you do, it won't help you follow your roadmap to success. If you're a present-dweller, make sure you're not running from the future. If you are or might be, write down a thorough description of where you want to be and what you want to have achieved fifteen or twenty years down the road. Then review it and see if you can spot any fears or events you're avoiding. If you do, reframe them by paying more attention to the upsides and then focus on those positives. It shouldn't take much coaxing to push your point of view a little further out.

The most empowering form of temporal thought is "future-think." People who think in the future are planners and often the most positive and upbeat in society. When faced with a particular situation, do you determine a course of action based on precedent? Do you make a snap decision that suits the moment? Do you play out the possibilities and devise a course of action aimed at realizing the best possible outcome? Whichever the case, as you strive for success, you should focus on the future because that's where your personal achievement resides. Thinking about the future is the first step of planning for it. By understanding where you're heading, you're giving your brain the guidance it needs in a tone that it comprehends.

To live more in the future, use your imagination to develop mental representations of how you want and need your future to look. Make sure the representations contain all the successes you crave. It's always better to dream big, without limits, than to follow a route outlined by your past. If you do nothing about what you want or need, you'll end up wasting your life by living with a view that controlled you rather than you controlling it.

That may not be as simple as it sounds, particularly if you're surrounded by people who live for the moment or dwell in the past. When you find yourself surrounded by those types, steer the conversation forward. Remember, your future starts now!

REMEMBER . . .

* Don't let your past determine your future.
* When you think of the past, always address the future to balance your mindset.
* Thinking about the future is the first step of planning for it.
* Success always lies in the future, and the future always starts now.

⟨23⟩

Fear of Failure or Success

FEAR usually prevents people from achieving the success they desire. It can take a number of forms, ranging from fear of failure to fear of success. In this chapter, we're going to discuss where these types of fear stem from and how best to disable them.

In order to succeed, your desire for success has to outweigh your fear of failure. Fear is both a noun and a verb, a state as well as an action. As the former, it means "an unpleasant emotion caused by the belief that someone or something is dangerous, likely to cause pain, or a threat." As the latter, it means "to be afraid of someone or something as likely to be dangerous, painful, or threatening." For our purposes, we'll take fear to mean avoiding perceived pain.

If you fear setting out to achieve success, somewhere in your mind, you're trying to avoid some form of pain. But what? What's stopping you? Here are a few likely candidates.

The human brain has a strange way of fearing failure. It classifies shame and embarrassment as forms of public humiliation. After all, you can't feel ashamed or embarrassed unless you think that others will reject you for doing something accidentally (embarrassment) or intentionally (shame). If you suffer from that type of fear, remind yourself again that failure doesn't exist—just a different outcome than intended. Also remind yourself that's it's always better to have tried than to go through life wondering what would've happened if you had tried. Remember the famous couplet from Tennyson's "In Memoriam":

'Tis better to have loved and lost
Than never to have loved at all.

A second common form of fear that prevents action is fear of not being capable of handling the success you seek when it arrives. Although there's some logic in that train of thought, your brain is misinterpreting the scenario. It's not success that your brain fears so much as mishandling success. *What would I do with all that money? What if I can't handle being CEO?* If you have that fear, you can send it packing by switching from the tangible aspects to the emotional reasons. For example, you want more money to help provide for and raise your family. Another version of this fear is a fear of disliking the results. *I don't want to be the boss, because I don't want to be the one that everyone fears.* This version can arise when you haven't thought all the way through the ramifications of your goal. Again, you can deal with this fear by focusing on the emotional benefits of your success. In this case, if you rise the right way, subordinates will respect rather than fear you. Once your brain recognizes that, it will feel better about achieving success than failing to do so.

We're susceptible to fearing the unfamiliar, a fear of the unknown, and of course we fear failure itself, one of the greatest barriers to success. We fear failing in business, so we don't start a business. We fear failing in a relationship, so we go through life alone.

But no great business tycoon, inventor, or lover who let his or her fears win out ever created anything. Think of some of the most successful men and women in recent memory: Richard Branson, James Dyson, Bill Gates, Steve Jobs, Anita Roddick, Sonia Sotomayor, Meg Whitman, Serena Williams. Each had to conquer serious fears on his or her way to achieving success. Can you imagine any of these esteemed entrepreneurs and champions backing away from those initial challenges just because he or she feared failing? I can't.

One of the greatest success stories in business history involved the creator of Kentucky Fried Chicken. In 1930, at the age of forty, Harland Sanders took a job at a filling station on the old Dixie Highway in North Corbin, Kentucky, eventually finding success by selling travelers the food he had learned to cook as a boy, including fried chicken. In 1955, as the interstate highway system took shape, the route for I-75 swung a two-mile drive west of North Corbin. Sanders's restaurant business depended on travelers, who he knew would switch to the new highway, so he knew he had to do something. Instead of turning to apathy or excuses, he set out to franchise his chicken recipe and then his restaurant. By 1963, Kentucky Fried Chicken, as it had been renamed, had become the biggest fast-food chain in the country, with over six hundred restaurants. The next year, he sold his company to a group of investors for a finger-lickin' $2 million and a lifetime salary. In 2013, the company pulled in $23 billion in revenue.

Take another example: In 1954, a young Elvis Presley played his one and only show at the Grand Ole Opry, showcasing the new rockabilly sound rather than the traditional country style that the audience was expecting. Afterward, Jim Denny, manager of the venue, famously told Presley, "You ain't going nowhere, son. You ought to go back to driving a truck."

Obviously neither man accepted the setback, and that's a good message for us all: If something frightens you, identify the source of your fear, and tackle it head-on. Never stop believing in yourself.

———

We've identified the different kinds of fear that can stand in our way. Now let's look at how to eliminate those fears.

The first step of removing what's holding us back is understanding the fear itself. Many of the root causes of adults fearing failure stem from childhood. In many cases, parents, teachers, relatives, and even friends may have undermined us when we were young. No one soon forgets that debilitating feeling of humiliation. In fact, it can stay with a person for the rest of his or her life. So let's disarm and overcome the issue.

As children, our fears are magnified. Everything seems larger to children because they're so small in relationship to their environment. If you don't have kids of your own, go back to your grade school, and see how amazingly small everything is when compared to your memory of it from decades ago. The same mental distortion also applies to how we remember moods and feelings. If as a child you were embarrassed by forgetting your lines in the school nativity play and the teacher chastised you publicly, the negativity of that event probably remains with you to this day.

To disarm it, look at the issue objectively. A seven-year-old forgetting what the three kings brought to Bethlehem while standing before a hundred anxious moms and dads isn't the end of the world. In the scope of the universe, it's a miniscule flub—so forget about it. After all, it's not as if you were president of the United States and forgot the words to the national anthem. Use the screen technique we learned in chapter 19 to crush and eliminate those bad memories.

A disabling fear of failure might derive from an event so traumatic that it continues to haunt you to this day. Even if you don't realize it, your brain has been protecting you from potential repeat trauma ever since. If you didn't know that was happening, that shows once more how our brains try to protect us, even though they comprehend only the emotional side of the story.

If fear of failure is holding you back, look for the symptoms. Are you reluctant to try new things or challenge yourself emotionally? Do you half-heartedly try something new only to give up at the first obstacle? Do you accept a challenge only if you're certain you'll accomplish it quickly and easily? If so, get ready for a change. Here are more tools to help you break free from the chains of fear.

Exercise

Conquer Your Fear

If you're afraid to undertake an activity, ask yourself what the benefits are of rising to the challenge. Think hard, be objective, and write them down. Don't let negativity or procrastination hold you back. Imagine how much better it would feel to remove the word *could* from your future. Change *I could have made it* to *I have made it*.

Another word to erase from how you face your fears is *but*. It's three letters long, *but* it can prevent anyone from taking action and achieving success.

Next, think through all the possible outcomes of taking the action you're contemplating, and write them down. Many of us fear failure because of that fear of the unknown. By removing the unknown, you disable that aspect of the fear. Be tough here. Look at the worst-case scenario, too. For each possible outcome, good or bad, identify how you would react to it. Often, you'll discover that it's not half as bad as you first thought.

As a final technique for overcoming your fear of failure, as we've been doing elsewhere, reduce the challenge into smaller chunks. If you want to conquer your fear of heights, take only two steps up a ladder. Once you become accustomed to that and your brain realizes you're still alive, go for one more. In other words, don't focus on the big-dream picture of running the company, getting straight As, or being happily married with 2.5 children. Pay attention to the small steps that will get you to your goals. They're much less daunting and easier to attain.

REMEMBER . . .

* To succeed, your desire for success has to be greater than your fear of failure.
* It's better to have tried and not succeeded than not to have tried.

* Never stop believing in yourself.
* Look at past failures objectively.
* To conquer fear of the unknown, plot out all possible results of doing something that scares you.
* Reduce challenges into smaller chunks to make them easier to achieve.

24

Imposter Syndrome

A NOTHER fundamental obstacle that often prevents people from succeeding is that they don't believe they deserve success.

Throughout life you'll meet scores of people who have great ideas but never get them off the drawing board. These people invariably have lots of sound reasons for not taking those first steps toward making their ideas reality. As you interact with them, notice how the focus falls so often on why not to take action as opposed to a more positive alternative. This powerful mental defense mechanism is hard at work in many of us, and it's only when we recognize its existence and limiting powers that we can take action to reduce its impact. Only when you recognize that your brain may be defending you for the wrong reasons can you give it a fresh set of instructions to do what you really want it to do.

It's unfortunate but true that many people don't believe in themselves. Perhaps they believe they don't deserve to be loved, don't deserve the emotional benefits that financial independence brings, don't deserve the chance to have a good job. In extreme cases, this mindset is called imposter syndrome. Those who have it believe, despite proof of their success, that they are frauds and that any achievements they have attained have come about by dumb luck, fortunate timing, or the like. But people who don't believe they deserve the success that they earn are sending a message to their brains that says *Don't succeed!*

If you entertain such thoughts, know this: They're nothing more than your brain giving you an easy reason for not trying. It's an excuse based on the misguided notion that you're unworthy. It comes from a lack of self-belief which derives, in turn, from the brain initiating its pain-avoidance mode. Your brain is trying to protect you from the hurt and pain associated with failure. It wants to shield you from the embarrassment, the ignominy, the emotional hardship of failure.

But I'm here to tell you it's a lie. Your brain isn't protecting you at all but rather preventing you from attaining success. The older part of your brain that protects you and manages your emotions is pushing a negativity feedback loop into the newer, decision-making part—and winning. Your emotional brain is preventing you from achieving the success you desire by shifting your focus from why you should succeed to why you shouldn't. Don't let your brain sucker you in to believing it.

If your brain continually sabotages your road to success, you need to change your mindset. Recognize that your lack of self-belief and overall feeling of not deserving success stem from your brain trying to protect you from fictitious pain. Only then can alter your outlook. Once you recognize that your brain is holding you back, you'll immediately recognize an alternative viewpoint. You do, indeed, deserve success and will achieve it! Fake it 'til you make it, as the saying goes.

But you can't win the battle so easily if you're surrounded by people who have a negative view of your finding the love of your life, losing weight, making a fortune, or getting that promotion at work. Naysayers and doom mongers will try to convince you that "You ain't going nowhere, son." And they'll do it every chance they get. So whatever you do, don't let friends, colleagues, or even family members derail you. People who are overly negative about the hopes and dreams of others are often projecting their own lack of self-belief. If they don't deserve success, nobody else does, either.

But what if you're not surrounded by a bunch of negative nellies? What if the problem is strictly yours alone? If that's the case, here's a technique you can use to chase those demons from your brain.

Love Thyself

Find a photo of yourself as a young child. Hold it in front of you, and look at it closely. Focus on the younger version of yourself, and ask, "What does this innocent, young child deserve? Doesn't she deserve the absolute best that life can offer? Even if she missed out years ago, shouldn't she have another chance today?"

With all the meaning and feeling you can summon, look at the picture, and say, "I love you." Then consider the question "Do I deserve success?"

Your own self-image isn't the only thing that can help spur us to success. Just as others can talk us into a state of negativity, so can they help us achieve more success—both actively and passively—and much more quickly.

REMEMBER . . .

* Thoughts about not deserving success are an excuse for not trying.
* Don't let your brain sucker you into believing you won't succeed.
* Replace mental negativity with positive alternatives.
* Avoid naysayers and pessimists who are projecting their own self-doubt onto you.

Part Three

MASTER COMMUNICATION AND RELATIONSHIPS

✳

25

Social Animals

SELF-PRESERVATION and reproduction are biological imperatives, but life doesn't consist merely of a vicious, solitary struggle for resources and mating partners. We rely on social interaction and cooperation to get what we need. Social interaction provides many benefits both to animals and humans. Studies have shown that whales and wolves are more successful in finding food if they hunt in packs. Similarly, our human ancestors tended to hunt in groups. We human beings have developed in such a way that functioning in groups is not only advantageous but also essential for the survival both of the individual and the species.

Michael Tomasello of the Max Planck Institute for Evolutionary Anthropology has shown that humans are hardwired to band together to form groups. We possess a neural capacity for empathy that influences us from the day we are born. For instance, babies cry at the sound of another baby crying. This shows how we empathize with one another's emotional needs even before we can feed ourselves, indicating just how important social interaction is to us.

Let's look at what happens when more than a few people gather in the same place at the same time. That's when the psychology of crowds comes into play, meaning the ways people behave in groups to ensure their survival. Most of us do so in culturally acceptable ways—although some of us notably don't. This conditioning derives from a number of fundamental instincts, such as withdrawing from danger, seeking nutrition, and craving friendship and love.

We all know that we tend to behave differently in groups than when alone. We laugh more at jokes in a comedy club than we would watching the same performance on TV. We cry more at tearjerkers when we see them in a movie theater. We dance more energetically at nightclubs than in our own homes. Our emotions, thoughts, and feelings drive our behavior as well as the behavior of those around us. So why do we behave more extremely in groups? Emotions are contagious. If we see someone smile, we're likely to smile ourselves.

We act, feel, and behave more openly in a group for another reason, which once again, comes about from evolution. Most emotions and feelings predate language, so as a result we can emote and feel more than we can verbally explain our emotions and feelings.

In our pre-language, hunter-gatherer days, our ancestors traveled in groups for safety because the eyes and ears of many proved more effective than those of just one. Groups could spot threats and potential food sooner and more effectively. But how did that happen before language came long? The answer: emotions.

Facial Expressions, the Fastest Communicator

Our brains have a mechanism designed solely for reading facial expressions, the fusiform gyrus. It sends signals to the limbic system structures, which contain the early warning system of the brain. This system switches our defense mechanisms on and off in a fraction of a second. A look of fear, terror, or alarm can alert an entire group in a split second, whereas language, which uses more recently evolved parts of the brain, can't communicate anywhere nearly as fast.

This alarm system alerts those around us through emotional signals, including facial expressions. We're conditioned by nature to pick up strong emotions and pass them along. To do so, we constantly watch for other people's posture, body language, and facial expressions, just as they do with us.

By paying attention to your own emotional signals, you can avoid inadvertently modeling yourself after someone else's physiology. Instead, you can take the lead by adopting the physical attributes that communicate achievement. Sit, stand, and walk how someone in control of his or her own destiny does. When you convey a sense of confidence, those around you will model themselves after you, and before you know it, you'll be the man or woman of the hour and the life of the party. Walk the walk and talk the talk, and others soon will follow.

Whenever you're in the company of others, your face and body are constantly communicating your perception of the situation, as are theirs. So you need to ensure that others aren't derailing your efforts with their own lack of motivation. Take a moment to examine the company you keep to see if the people around you are creating roadblocks to your success. They may be doing so without even realizing it, so you don't have to ostracize them from your life automatically. Projecting positivity yourself will pick them up and reinforce your own rewired positivity.

Yet some people still feel uncomfortable and ill at ease in groups. But if the group doesn't lie at the heart of that discomfort, what else could it be? Once again, the answer often lies deep within our own imperfect brains.

REMEMBER . . .

* We're programmed to form and interact in groups.
* Emotions are contagious.
* We communicate through emotions faster than we communicate through language.

{26}

Evolution and Attraction

WE know that in only a fraction of a second after meeting, people make initial judgments about others. But what exactly is it that people find attractive in others? As we've seen, material possessions can play a part, but our instincts determine beauty in a way used by our ancestors long before bespoke suits, designer watches, expensive cars, luxury handbags, or even boob jobs, cosmetics, face-lifts, or liposuction ever came along.

Most men will turn to look at a young woman in a short skirt, and some women will give more than a passing glance to a man driving a Mercedes convertible. But we're all susceptible to certain physical traits that make some humans look immediately more desirable than others. Much of the reason for this lies in our evolutionary hardwiring. Although men and women have different preferences in terms of what they find attractive in the opposite sex, many of the traits we're attracted to mirror those of other species in the animal kingdom. We share these preferences for a single reason: reproduction. Healthy heterosexual males and females, both human and animal, are programmed to produce the next generation of their species.

For the moment, let's put aside the romance of courtship. Our brains have evolved to ensure our survival. That means we are attracted to and seek out aspects in others that make them good investments from a reproductive point of view. Scientists have discovered that when we see someone whom we find attractive, some of the older parts of our brains light up. That neural activity leads to an emotional response that includes mental and physical sensations such as excitement, nervousness, racing hearts, and sweaty palms.

Men are attracted to younger women because they have a greater likelihood of being able to bear children. Men tend to look for womanly attributes that signal youthfulness: thick, lustrous hair; clear, smooth skin; bright eyes; and full lips. Notice how many of the beauty products in stores target exactly these aspects that men instinctively find so attractive. It's no coincidence. Men also pick up on certain aspects of behavior, including facial expressions that communicate youth and vigor (lots of animation) and a style of walking that signifies health and vitality (a spring in the step).

Women optimize their attractiveness to men in varying ways, include wearing high heels, which give them the appearance of increased height and longer legs. High heels also have the effect of forcing women's backs to arch, pushing the chest forward and the buttocks back. In other words, as fashion historian Caroline Cox writes, "Men like an exaggerated female form." If a woman doesn't wear high heels, she generally will appear less attractive by comparison because others are still wearing them, presenting a catch-22. Women who want to find a mate must continue wearing heels because other women who want to find a mate are wearing them.

Women also exaggerate their appearance through cosmetics: evening skin tone, making the eyes look larger, and presenting fuller, redder lips. An equally important part of this physical presentation, according to many sources, is that it makes the wearer look as if she is at her most fertile and most likely to become pregnant.

Researchers have discovered that a woman's scent is also an important attractiveness factor for men to consider. At certain times during a woman's menstrual cycle, her scent is more appealing to men due to the varying levels of estrogen that the female body produces each month. Researchers from Charles University in Prague have identified the times during the menstrual cycle when women smell their most and least attractive to males, and the answer might surprise you: They smell most attractive between the first day of menstruation and ovulation, roughly the first twelve days of their cycle. So before smothering yourself in perfumes and body scents, think about whether you're hiding something that potential partners will find instinctively attractive.

But artificially exaggerated features aren't just a woman's game. Women are attracted to men older than they are because they have a greater likelihood of being able to provide for a family. Women look for manly attributes that signal strength and viability: thick, lustrous hair; clear, smooth skin; a strong jaw line; muscular development; and height. Men also have a whole host of tricks to look more attractive to women. They work out to become physically stronger, they wear colognes and other fragrances that emphasize the presence of testosterone, they remove or groom their body hair to help showcase symmetry—a visual representation of genetic stability, and they wear shoes that increase their perceived height.

Women recognize that mating is a massive commitment that includes both pregnancy and also rearing children, so another part of what women attractive in men therefore involves security and reliability. They instinctively look for men who can provide food, shelter, and other resources for both mother and offspring. That's another reason that women find athletic-looking men appealing: they're more likely to be good hunters and offer physical protection. Women are attracted to successful males because they perceive them as better able to provide for a family's future, and they like generous men because generosity gives an indication of how they may behave in the future.

As result of all of the above, any advertiser worth his or her salt knows and has exploited the fact that sex sells. That lipstick or mascara will make you more alluring. That wrinkle cream will make you look younger. That body spray will make you smell like an athlete. That car will transform even the dullest and puniest man into an instant god—or so the advertisers want us to believe.

But what advertisers and magazine editors say constitutes the ideal look doesn't always align with what the opposite sex is seeking. Remember, women's magazines are selling products *to women*, so they're exploiting how women think they themselves should appear rather than what men actually want. The same goes for men and men's magazines. Most men aren't attracted to models so thin that they look malnourished, and most women aren't attracted to men so muscled that they could win a bodybuilding competition. Keep that in mind the next time you watch television or look at a magazine rack.

Sex sells, it's true. But the hardest sell in the world can't hold a candle to human evolution and the hardwired demands that our brains place on our quest to find the right mate. Fortunately we can make the most of what we have. We're not stuck with the hand that genetics dealt us. All of us can increase our sexual desirability quotient. Let's take a quick look at what we can do to maximize our looks by making minimal, easily achievable changes.

Clear skin conveys youth and health. Are you treating your skin well? Religiously follow a routine of cleansing and moisturizing—men and women alike. You're aiming for a clear, consistent complexion. To that end, avoid excessive sun exposure. A light tan may look good, but it isn't healthy for the skin. According to the US Department of Health and Human Services, "There is no such thing as a safe tan. The increase in skin pigment, called melanin, which causes the tan color change in your skin is a sign of damage." Skin cancer is the most prevalent form of the disease, and it's also one of the most preventable. If you do want to hit the beach, always wear plenty of sunscreen.

Your smile is one of your best attributes, and it says a lot about you evolutionarily. Are your teeth symmetrical, strong, and bright? Take care of them by brushing and flossing regularly and visiting your dentist twice a year for regular check-ups and cleanings. Your smile doesn't stop at your teeth, though. Your lips offer one of the first signs of dehydration. In the wrong light, chapped lips can appear at best unsightly and at worst like symptoms of an STD. If you suffer from chapped lips, drink six to eight cups of water a day, and lock in that water with a lip balm that contains little to no alcohol (which dries out the skin and makes you apply more). A light dab of Vaseline right before a big date can be a lifesaver.

Putting a little effort into styling your hair can increase your attractiveness while boosting your confidence. Get a haircut whenever you feel your hair is getting too long or shaggy, every month or two. Pay attention to how the hair stylist or barber styles your hair and try to do the same at home—but don't over do it! Fried locks and overdone coifs won't make you look more attractive, but instead they'll make you look like you either don't know what you're doing or you're trying too hard. Check your hair a couple of times during the day to make sure it's still doing what you want it to do.

Women can fake the art of looking good with a little effectively applied makeup. You might not *need* makeup to look attractive, but it can significantly alter your appearance when you want it to. Remember that less is more. (Bridal makeup and stage makeup are meant to make features distinct at a distance, not for everyday, close-up encounters.) Too much, and it will look like you're trying to hide something.

Men, too, need to pay attention to details. Make sure your fingernails are clean and clipped, and remove excess hair from your ears, nose, and eyebrows. That same cleanliness should extend to your clothing, both outerwear and underwear for obvious reasons! Also, for men in particular, make sure you smell nice. Women have a much more acute sense of smell than men, so wear antiperspirant or deodorant and a *hint* of cologne or body spray. Again, less is more.

Now that we've ticked off the basic elements of attractiveness, in the next chapter we will look at how you can appear more approachable. That boils down to not only using what you already have but also using it more effectively.

REMEMBER . . .

* Men seek attributes in potential mates that signal youthfulness and fertility.
* Women seek potential mates who display signs of strength and capability.
* Women naturally smell more attractive during the first twelve days of their menstrual cycle.
* Women have a stronger sense of smell than men.
* Sex sells, but don't let advertisers fool you about what the opposite gender wants.
* Pay regular (but not excessive) attention to your skin, smile, and hair.

⟨27⟩

First Impressions

IMAGINE walking into a room full of people. A number of them glance in your direction. On average, how long do you think it takes their brains to develop a meaningful first impression of you? Scientists have quantified how quickly we arrive at that first impression: one hundredth of one second!

That's the power you have between your own ears, and once again evolution explains why we make such rapid judgment calls. In our early history as a species, when one individual spotted another, he or she needed to know quickly if the second person was a potential friend or foe. The faster the processing time, the better equipped we were at navigating various social situations. As a result, humans developed an efficient and extremely rapid way of looking at different features on another person and deciding what sort of individual he or she is likely to be. Our brains pay particular attention to the face, focusing in on the eyes and the shape of the mouth and analyzing facial expressions. In this hundredth of a second, our brains assess up to ten thousand facial expressions—the number that each able-faced person can make—giving us an early warning system for fight or flight so that our bodies prepare instinctively for the optimal course of action.

Cartoonists exploit our ability to judge faces in their creation of caricatures. Good characters tend to have big eyes and higher eyebrows, while the bad guys have smaller eyes and lower eyebrows. That's one pair of examples, but your brain can detect a huge number of aspects in another's face, and it follows that others are judging you in the same manner. First impressions really do count, so it's up to you to manage and develop the impression you give for maximum personal achievement. According to Roger Ailes of *SUCCESS* magazine:

> You've got just seven seconds before others will have formed a considered opinion about you: Considered

opinion consisting of the instinctive first impression combined with an initial cognitive evaluation. As soon as you make your entrance, you broadcast verbal and non-verbal signals that determine how others see you. In business for example, those crucial first seven seconds can decide whether you will win that new account, get that additional financing, or succeed in a tense negotiation.

Fortunately, you can use science to refine a person's initial impression of you. How? Stand in front of a mirror and give yourself a visual audit: What do others see, and is it what you want them to see? Think about what how you judge other people, and you'll recognize what you communicate to others, both from an evolutionary perspective and a more modern-day standpoint.

For example, what does your posture say? Do you stand upright or slouch? Are your shoulders square or rounded? What about your face? People especially look at the eyes and mouth (more about those later). How are you dressed, and what message does that send? It's not so much the individual components by which others judge us, as it is the combination of all the parts. When you ask yourself these questions, ask them aloud and answer them aloud as well to help you see yourself more objectively.

<hr>

Exercise

Take Command of Your Physiology

You can get your body to appear the way you want without taking acting classes or drugs. Here are a few simple steps you can take to create a good first impression:

1. When you're about to go on a job interview or to a business meeting, stare at something bright immediately before entering the room. When you look away, your pupils will

enlarge as they adjust to the lower level of light in the room, which will convey a sense of honesty and interest.

2. If you flush when nervous or embarrassed, think of another part of your body. As we learned earlier, your brain will divert more blood to that body part, thus reducing the redness of your face.

3. When approaching someone new from a distance, flashing a quick smile will make you seem more approachable because you're showing that you are not a potential social threat.

Our cultural dress codes also say a lot about us. Designers and other creative people tend to dress differently when appearing before corporate board members, for example, than they do day-to-day. Ask yourself how you should look in different scenarios, whether for a job interview or first date. What do you want to convey? Straight-laced and serious, imaginative and unconventional, comfortable and relaxed? In structured gatherings like office meetings, birthday parties, and weddings, take note of who looks best in such situations. Endeavor to understand what it is about their appearance that makes it and them so appealing.

You should create and develop the first impression people have of you based on three characteristics: posture, overall appearance, and face. Posture relates to how you carry yourself; it's your deportment. Appearance concerns what you wear, how you wear it, general cleanliness, and so forth. It also refers to whether you're tall, short, fat, or thin. Not surprisingly, your face is the most important aspect of all.

Put Your Best Foot Forward

Here are pointers on how to look your best for that all-important hundredth of a second.

First, consider how you want to be perceived. Different situations require that others perceive you in different ways. A job interview is a very different scenario than a blind date. Once you've determined the image you want to convey, begin creating the perfect you from the standpoint of first impressions. Physically and mentally copy others whom you admire in the ways that you're trying to project. In psychology, this is called modeling. Once you identify somebody whom you'd like to emulate, answer these questions and try to model them: What are the behavioral patterns of that person? Why would he or she be successful in this scenario? What's the visual difference between that person and me?

Remember that the most communicative part of your appearance is your face, so you need to pay particular attention to what it conveys. Avoid covering your eyes if you want to convey openness and trust. Sunglasses may look cool and help you see more clearly, but they can also look like you're trying to hide something. For men, make sure your facial hair isn't projecting the wrong image by hiding your mouth. For women, check your makeup. A recent study revealed that when women look at other women and identify what they believe to be the ideal amount of makeup either for a job interview or a night out, they misjudge the amount they need to apply to their own faces when they try to achieve that same level for those situations. In the study, the research subjects applied 50 percent more makeup to themselves than on the images they had identified as ideal, but they thought they looked the same. Honest second opinions are important for good first impressions.

Here's another piece of advice. The commercial worlds of advertising and communications use highly skilled copywriters to make sure messages come across in just the right way. They also use expert art directors to ensure that photography and other imagery conform and convey just the right mood, style, and emotion. But look at most people's online dating profiles, résumés, or blog profiles, and the clearest message is often *This will have to do*. A hasty description below a poorly selected photo is worse than a good photo and no description at all.

First impressions count, and people can form them before you even meet. Once you understand what you see in others and how that creates your impression of them, you'll better know how to hone your own projected image. Others are going to judge you, so make sure they see you in your best light, both in terms of how you describe and portray yourself. You want people to remember you for the right reasons!

REMEMBER . . .

* Give yourself a visual audit to determine what impression you're giving others.
* To create a good first impression, focus on your posture, overall appearance, and face.
* Before a job interview or important meeting, stare at something bright immediately before entering the room.
* Model yourself on others you admire, but get a candid second opinion in order to make a good first impression.
* Take care that sunglasses, excessive makeup, or facial hair isn't concealing too much of your face so you can convey a sense of trust and openness.
* First impressions count, so take the time to craft the right text and select the right images for online dating profiles and résumés.

28

The Second Moment of Truth

OK, you're sitting in a room, face-to-face with another person. Each of you has formed some initial impressions about the other. What next?

Well, first of all, you're a trained people watcher now, so recognize that the chances of your instincts being right are pretty good. But what that other person thinks of you may fall somewhat wide of the mark. Fortunately, you can still alter his or her perception of you by working on the second moment of truth.

Let's say that someone initially perceived you as shy and aloof because you were looking down as you joined a party. You have a second chance to amend that perception with a little fine-tuning. Try adopting this tactic that politicians use: Look into the distance and pretend to acknowledge somebody you know with a flamboyant "Hi, good to see you!" gesture. You're not focusing on anyone in particular at all, but in an instant that gesture has helped erase that shy and aloof label. You might not be the life of the party (yet), but at least you won't seem like you don't know anyone or don't want to be there.

Another example: Imagine that someone perceives you as being less than honest because you entered a bright room wearing sunglasses. After you've removed your shades, concentrate on making direct eye contact with others and go for some form of acknowledgement, such as a subtle nod of the head. Another example plucked from the politician playbook is you donned a nice suit and tie, but after arriving at the party you find that you're overdressed for the occasion. To use the opportunity of that second moment of truth, steal away for a moment and take off your tie, unbutton your collar, remove your jacket, and roll up your sleeves. Others will revise their initial perception that you're stodgy or unapproachable and perceive you as more down-to-earth and congenial.

Once you perfect the practical ability to sway people's judgments of you, you'll be better able to use it as a tool to deal with others. Let's say a center-of-attention egomaniac at that same party made you feel insecure as you entered a room, so you instinctively looked down. When joining the conversation, look to a point above his eyes, at either his forehead or hairline, and you'll switch the power dynamic so that he no longer feels in charge. Once you win that opening salvo, you can adjust your mannerisms to shape his behavior toward you. Another way you can go toe-to toe with a group of headstrong people is to position yourself centrally within a group. Don't cower on the edge of the gathering, but get in the middle and make yourself a central player. Being surrounded by other people will give you more confidence and impact their perceptions so they perceive you as on center stage.

The second moment of truth works equally well in romantic situations. Suppose that you accidentally sent the wrong signal to that hot guy or gal standing in the corner and came off as snobbishly uninterested. You can win his or her attention back by making eye contact for longer than necessary and flashing that ever-important quick smile. You'll be making polite conversation before you know it. Conversely, you can send a loser packing by making quick eye contact and dismissively breaking it off. In some respects, women benefit from this flirting technique more than men because women tend to be more advanced at making and reading social cues. If a man makes a bad impression in that critical hundredth of a second, there's not much he can do other than grin sheepishly and hope the woman gives him a second chance. (Thankfully women are also more likely to pick up and recognize an apologetic look!)

Invasive Influence

Here's how to alter how those around you feel, perceive you, and react to your presence. Before unleashing it on the world at large, though, first try this technique with friends and colleagues because it can be remarkably powerful and persuasive.

1. Sit next to someone you know at a bar, in a restaurant, or at an office desk. As you talk, move items such as glasses, cutlery, or stationery slowly toward him or her—only about an inch at a time. Do it subtly so that he or she doesn't immediately notice your advances.

2. After a while you'll notice changes in the other person's physiology. As he or she either backs away or pushes the items back toward you, note posture. Is she attacking? Is he becoming defensive?

3. Now, try the opposite direction. Pull those items back toward you. You've now got a great, subtle tool for when you want to influence how a person next to you feels. The other person will relax physically. For example, if somebody near you is becoming a little too overbearing or intimidating, inch objects toward him or her. Subtly invading the other person's space will unsettle him or her and reduce that person's level of confidence. Use that technique when negotiating and you want to appear more assertive. On the other hand, if you're on a date and want the other person to engage a bit more with you, move items toward yourself. That communicates that you're comfortable with him or her coming closer and encroaching on your space. The same technique in reverse can signal discreetly for someone to back off. Whatever you want to achieve using this technique, the best part is that you don't have to resort to words to express yourself.

Appearance, posture, and body language all translate in the first few seconds of meeting someone into a first impression. That first impression is extremely important, but you always have a second chance to revise others' impressions of you. Use the opportunity.

REMEMBER . . .

* You always have a second moment of truth to correct a misguided or accidental first impression.
* The second moment of truth can help you manage other people and their personalities.

29

A Beginner's Guide to Body Language

THE phrase seems self-explanatory—it's the unspoken language that one body uses to communicate with others. But is it that simple?

Body language is the physical means by which humans communicate nonverbally. We do this by way of posture, gestures, facial expressions, and even the ways in which we move our eyes. We continuously send, receive, and interpret such signals both consciously and subliminally. Our material possessions affect how others see us, but much of what people think of us comes from how we look and behave. Sometimes we subconsciously exhibit our emotions and feelings, but thankfully through awareness we can always choose how to appear. This allows us to improve others' perceptions of us without spending a cent. Plus, you'll find that using what you already have is far more effective than undergoing expensive surgery or buying costly balms and miracle cures to attain something unreachable. Develop what you have before attempting to acquire something you don't. You'll become happier, more positive, and more goal oriented.

Researchers suggest that 55 to 70 percent of all human-to-human communication is nonverbal, the percentage depending on the subject matter. Directing somebody from point A to point B differs from telegraphing anguish or implying ennui to another person. When you study the body language of others, you'll discover clues about their attitudes and states of mind. These clues may indicate aggression, attraction, boredom, displeasure, or any one of many other states. A powerful communication tool, body language not only helps people communicate more effectively, but it also prevents unintentional gestures that might leave a negative impression. Covering your mouth when you speak, slouching, or crossing your arms convey meaningful nonverbal information. In poker, for example, such inadvertent gestures are called "tells" and prove very useful for detecting who has the upper hand.

The study of body language—known as *kinesics*—has become a science in its own right with numerous subcategories, including haptics (the study of touching or physical contact), oculesics (the study of eye movement), and proxemics (the study of personal spatial relationships). But a world of difference lies between the science of body language and how the average person uses or even understands it. Some of us can read body language fluently, while others remain completely oblivious to it. But with a little extra effort, almost anyone can learn to interpret nonverbal cues, and with practice it can become second nature.

Pay attention to how close to you people sit or stand. The closer they are, the more attracted they are. The farther away, the less they care for you. If you move closer to somebody, and he or she moves farther away, take that as a sign that he or she doesn't want your relationship to become any more personal than it already is. If the other person doesn't move away, he or she is signaling receptiveness to you. If that person counter-responds by moving closer still, he or she is signaling mutual affection and comfort.

Watch the head position of the people with whom you communicate. When we tilt our heads, we are signaling either sympathy or, if smiling simultaneously, flirtatiousness. But a cocked head—tilted sideways and turned to the right or left—indicates confusion. Think of how a dog cocks its head when it hears a funny sound. A bowed head may indicate suppression or avoidance of something. If you lower your head when complimented, you're signaling shame, shyness, or unworthiness, or you may be showing submissiveness or deference.

Looking at the eyes of those with whom you are communicating can reveal a lot about what and how the other person is thinking. People who look to the side can be nervous, lying, or distracted. People who look away from you as you speak could be signaling submissiveness. But looking up and left might indicate the recollection of a memory. If someone looks down at the floor a lot, he or she is probably shy or timid. People also tend to look down when upset or trying to hide something. The act of looking down can cause you to feel down, so when you're feeling blue look up to reduce the intensity of that lowly feeling.

People who study kinesics know the importance of mirroring. If someone mirrors or mimics your appearance, it's a genuine sign that he or she is interested in you and wants to establish a rapport. You can test another person's feelings while talking to him or her by changing your body position to see if the other person follows suit. You can also use mirroring when you want somebody to think you have an affinity for him or her. But keep it subtle, and don't make it look too artificial. For example, adopting a similar sitting posture as the person interviewing you for a job or to whom you're selling something will endear her or him to you—so long as it's not exaggerated enough to make the other person think it's an act of mockery!

Pay attention to arms and hands, too, since they reveal a lot about a person's state of mind. People with crossed arms are not-so-subtly closing themselves to social influence or interaction. They're literally putting a barrier between themselves and whatever's in front of them. Some people cross their arms as a habit, but doing so may indicate that the person is uncomfortable with his or her appearance or trying to conceal self-consciousness. If a person's arms are crossed while his or her feet are shoulder width apart or wider, he or she may be signaling toughness, authority, or antagonism.

If someone rests his hands behind his neck or head, he's open to discussion or relaxed. If her hands are on her hips, she might be waiting for your next move, or she might simply be impatient or tired. Closed or clenched hands might indicate anger, irritation, or nervousness. Clenched hands definitely signal tension of one kind or another, and a number of gestures can betray nervousness, so be mindful of them both in others and in yourself.

A word of caution, though: Be careful when interpreting body language not to miss the obvious while hunting for the subliminal. If someone brushes his hair back with his fingers, it might be that his convictions conflict with yours . . . or maybe he just needs a haircut. If he raises his eyebrows at the same time, though, you can bet that he disagrees with you. If the person is wearing glasses and she constantly pushes them onto the bridge of her nose with a slight frown, she might be signaling that she disagrees with you, but she also could just be adjusting poorly-fitting spectacles. Lowered eyebrows and squinting eyes illustrate an unsuccessful attempt to understand something—assuming the other person isn't trying to observe something far away. Always look for groups of clues, not isolated gestures incongruent with everything else being communicated.

The final areas of the body to heed are the legs and feet. Rapid toe tapping, a sudden shifting of weight, or an unnatural movement of the foot often communicate impatience, excitement, intimidation, or, again, nervousness. Foot tapping could be a sign of wanting to leave—or too much caffeine. Slow shuffling of the feet, on the other hand, indicates boredom—or physical fatigue. A person sitting with feet crossed at the ankles generally implies that she or he is relaxed and at ease. Standing with feet close together often means trying to behave correctly and fit in with the situation. Feet together may also mean that someone feels passive or even submissive toward you. Even the subtlest gestures can have profound meaning. People who point their feet in the direction they want to go are indicating a readiness to leave. If someone points his feet in your direction, it may mean he's interested in you. If his or her feet brush yours, pay attention to what comes next. If he or she reflexively apologizes, it means that footsie wasn't intended. If neither of you apologizes, get ready for action because you're both definitely interested in each other.

From an action standpoint, signal approachability by pointing your body, legs, or even a single foot toward whomever you want to focus on you. Doing so creates a subliminal message that grants a person permission to approach. Indicate that you're communicating honesty and openness by showing the undersides of your wrists and the palms of your hands, which indicates that you have nothing to hide. These examples demonstrate how easily we can alter our personas, and as we change how we look, our brains think differently. When you communicate that you're attractive, you feel more attractive, which leads to a more positive mindset—not to mention opening yourself to a more proactive social environment.

Pay attention to how actors act, and you'll acquire even more tools to use in your own life. In particular, watch how the same actor plays different roles. Daniel Craig plays suave, international-spy James Bond one way, but he portrays the unstable Connor Rooney in *Road to Perdition* quite differently. The more you recognize how others use their bodies to convey emotions, the greater your arsenal of communication tools.

But again, proceed with caution. Yes, as much as 55 to 70 percent of what we communicate happens without any verbalization, but you can increase your chances of understanding someone or projecting emotion nonverbally by considering *all* the little clues of an interaction. As you do, embracing the basics of body language will help you understand people's feelings and, therefore, give you a powerful tool in becoming a more skilled communicator.

REMEMBER . . .

* Body language is how we communicate nonverbally.
* More than half of all human-to-human communication is nonverbal.
* The eyes reveal what and how the other person is thinking.
* Physical mirroring indicates a sympathetic state of mind.
* Be careful not to miss the obvious while hunting for the subliminal.
* Always look for groups of clues to avoid missing the obvious.
* When you project an emotion, you feel that emotion more deeply.
* Watch how actors act for more nonverbal communication tools.

⟨30⟩

What's in a Face?

LONG before we had language, we had faces, which sometimes communicate more than words ever can. Our emotional reactions show in our faces nearly instantaneously. This initial, instinctive reaction is revealing both when you see it on the face of another person and when he sees it on you. As with body language, evolution has driven or influenced most of our facial communications, so not surprisingly, much of what we know about how our faces communicate our feelings comes from the pioneering work of Charles Darwin.

Ever since *On the Origin of Species* published in 1859, scientific studies have been uncovering what our faces reveal. In Darwin's 1872 book, *The Expression of the Emotions in Man and Animals,* he identified human actions as strongly linked to animal behavior. These actions derive from our evolutionary needs. In the study, Darwin also recognized the universal nature of bodily expressions: "The young and the old of widely different races, both man and animals, express the same state of mind by the same bodily movements."

From the oldest evolutionary perspective, Darwin identified that human facial expressions evolved from body parts known as branchial arches. These are parts of a gill associated with the extraction of oxygen from water. That means that human facial expressions started developing before our evolutionary forebears even crawled from the oceans some three hundred million years ago.

Darwin also pointed out that happiness is visible from farther away than any other emotion. That's because we have evolved into social creatures who seek the company of groups with whom to interact. The appearance of happiness makes us more approachable, less threatening, and therefore more desirable as part of the group. Another example is the baring of teeth when angry, which stems from the threat of biting an opponent or predator. Snarling, which displays one or both canine teeth, therefore signals a preparedness to fight.

Humans indicate dislike or disgust by forming a unique downward shape with the mouth, which prevents anything from entering it while at the same time partially blocking the nostrils by pushing up the top lip. That expression easily instructs others to avoid eating food that smelled or tasted bad and was likely rotten.

Focusing on the eyes, Darwin again made an excellent link between emotional expressions and evolution. When a person partially closes his or her eyes to signal disdain, the emotional message is that whatever's in sight isn't worth beholding. When someone opens her eyes wide in awe, this facial act communicates that others need to see and understand more about what she sees. Lastly, when someone is acting—that is, faking a facial expression—Darwin stressed that the face moves asymmetrically, with one side moving more than the other. Conversely, most natural facial expressions occur symmetrically.

But the study of facial expressions has continued to develop in the years since the Victorian era. Carl Sagan, who was a professor at Cornell University and the renowned host and cowriter of the original *Cosmos* television series, hypothesized that, as a survival technique, human beings are hardwired from birth to identify the human face. That skill allows people to use only minimal details to recognize faces from a distance or in poor visibility. The evolutionary advantages of being able to discern friend from foe with split-second accuracy are obvious and numerous.

People quickly objectify faces, reducing them to their most fundamental elements, a few circles and a line. Recent studies help to explain why. The latest research by a team from Bangor University in the UK has discovered that, when recognizing three-dimensional shapes, people naturally focus first on areas where concave sections intersect with the surface of the overall shape. This predisposition toward the concave regions, as part of shape recognition, is linked to an instinct used for recognizing faces. In other words, objects resembling faces automatically activate our brains. Those objects trigger both an emotional state and a search for familiarity with the object. Interestingly, this occurs even before the conscious mind begins to process or receive the information.

It seems simple enough—spotting a face in a cloud overhead is literally child's play—but the complexity of the process runs much deeper than we realize. Paul Ekman, author of *What the Face Reveals* and a number of other studies, has developed the Facial Action Coding System (FACS) that classifies people's facial expressions as indications of their emotions. The face is the only place where the muscles connect directly to the skin. This, combined with knowing that emotional responses happen faster than rational thought, means that facial expressions often reveal more of the truth. But according to Ekman, we humans have some three thousand facial expressions directly related to our emotions, yet the English language has fewer than two hundred words to describe them.

Indeed, Albert Mehrabian—professor emeritus of psychology at UCLA and best known for his work on the importance of verbal and nonverbal messaging—created the 7-38-55 rule, often known as the three Vs (verbal, vocal, and visual). We'll look at it again in more detail later, but for our purposes right now the three numbers represent, in percentages, what constitutes our positive emotions toward someone. Words alone count for 7 percent, tone of voice gets 38 percent, and body language, including facial expressions, makes up the the lion's share at 55 percent.

Based on this and other science, it's no wonder that people spend more than $40 billion each year on cosmetics. As Dale Carnegie, author of *How to Win Friends and Influence People*, said, "The expression a woman wears on her face is far more important than the clothes she wears on her back." But the fact is, makeup works, and it does so because our bodies are programmed to perceive sexual signals from the coloration of people's faces. Makeup has been used for centuries in similar ways by differing and diverse cultures to color different areas of the face. We're drawn to made-up faces because they tap into our primal urge either to find or look like a young, healthy mate who will produce offspring thus passing on our genes.

As we've seen, although most women know the benefits of using makeup, not many are as good as they think they are at applying it. Many women—excluding actors and stage folk for the moment—apply too much. In a study conducted at Bangor University, women first looked at images of other women and chose from ten different levels of makeup which had the best makeup for either looking attractive or appearing more dominant. The second part of the study asked the participants to apply the same amount of makeup to themselves so that they looked either their most attractive or most dominant. The results showed that, on average, women applied 50 percent more makeup to themselves than necessary, making themselves look significantly less attractive. The problem wasn't how the makeup was applied but rather how much.

The angle of your face with regard to how you tilt it—chin up or down—strongly indicates how attractive people are likely to find you. Researchers at the University of Newcastle discovered that after showing participants several computer-generated male and female faces looking straight ahead but with varying head tilts. The results are significant. If a woman tilts her head back, men tend to find her less feminine and attractive. If her chin is down, she's perceived as more feminine and attractive. The same head tilts for men had the opposite effect, though. Men looked less masculine and attractive with heads tilted forward.

Another study, undertaken by the University of British Columbia, surveyed 1,084 heterosexual men and women and concluded that women find men less attractive when they smile compared to when they look swaggering or brooding. In contrast, the same study found that men find women more attractive when they do smile and less attractive when they look proud and defiant.

But something else caught my eye about the images in the study from the University of British Columbia. The faces that participants rated in terms of attractiveness included the same boy, girl, man, or woman in a number of different poses, exhibiting a variety of facial expressions. Each facial expression was posed or staged. A polite smile, known as a Duchenne smile, shows the usual shape of the mouth and a hint of teeth without any other facial activity—no crow's-feet or raising of the cheeks. Something wasn't right.

Studies conducted by the Institute of Emotions Research in the UK—a not-for-profit organization funded by my own company, Shopping Behavior Xplained Ltd.—and by Erin Heerey of Bangor University School of Psychology each identified a preference for a genuine smile over a staged smile. In one study, respondents were asked to choose a free sample of a dental product from one of two apparently identical display boxes in supermarkets. Both boxes had an image of the same young lady smiling on them, one genuine and the other staged. The study results showed that more participants instinctively chose the real-smile box.

This means that the study from University of British Columbia needs refining before it can be taken as gospel. But in the meantime, here's how you can tell when a person is expressing one or more of the primary emotions.

Facial Recognition

A person shows happiness by raising the corners of his lips in a smile. That part we all know, but if the happiness is genuine you'll notice the crow's-feet radiating from the outer edge of each eye. At the same time, the muscles in his cheeks also rise. If while playing cards one of your opponents looks at his hand and, before he can stop himself, gives one of these smiles, chances are he has a good hand. The key to reading happiness, then, is to decide whether a person's smile is genuine or merely polite. The former reveals truly how he feels, while the latter is concealing another emotion or even a lack of caring altogether.

When a person feels fear, his eyebrows rise while coming together somewhat. Simultaneously, his upper eyelids also rise. In the case of significant fear, the mouth becomes involved, too, as the corners stretch downward and back. Observing fear can tell you how you should relate to a person. When discussing possible activities for a second date, for example, if the other person expresses fear, ask yourself why. Is the fear about your suggestions or about dating you? If it's the latter, perhaps it's time to move on.

The telltale signs of sadness include the inner corners of the eyebrows going up while the outer corners of the mouth turn down. In a sales scenario, if you detect sadness in your customer, it's probably not the best time to go for the kill. Better to satisfy any objections or hesitations before closing the deal.

Surprise shows up on people's faces more than you'd expect. Sometimes it's simply because they didn't expect a particular question, in which case surprise gives way to another expression, depending on the question and the person. Surprise manifests itself with the entire face opening up. Eyebrows raise, eyes open wide, the mouth gapes. A person showing surprise is caught off balance and needs to decide on how to react. Watch for facial clues that follow surprise before you decide how or even whether to continue.

Disgust exhibits itself when the eyebrows pull downward, the nose wrinkles, and both sides of the upper lip rise. Disgust on the faces of those around you means all is not well. Proceed with whatever you're doing using greater than normal caution. Contempt, sometimes catalogued as a form of disgust, communicates an aggressive lack of respect or even disdain for something, and it's also easy to spot: one side of the upper lip rises.

Anger, the last of our primary emotional expressions, features lowered eyebrows and, often, raised upper eyelids.

In cases of extreme anger, the lower lids may also rise. Another telltale sign is the tightened lips, which appear thinner. In nearly any interpersonal setting, if somebody is exhibiting anger, you should try to find out why and diffuse the situation.

So how can you use this newfound knowledge of faces to further your own personal achievement and discover what your own face says about you? First, you need to understand a key characteristic about faces. We've been discussing facial expressions as a group, but they actually fall into two distinct categories: macro and micro. Macro expressions, which last from half a second to four seconds, often occur in our daily interactions with others. A smile from the mailman, a nod from the garage mechanic. Micro expressions, on the other hand, are subtler, lasting for less than half a second when a person is only beginning to feel an emotion. They sometimes take place when people try to conceal or repress their feelings. (Remember the face your card buddy made when he drew the winning hand or the look on your face when you got yet another pair of hand-knitted socks for your birthday.)

You can use this science of face watching to gain control over a particular social situation. To read another person's face accurately, you need to observe their expression before they respond verbally. Armed with that information, you can manage numerous interpersonal situations. You'll learn the chances of a person accepting your invitation for dinner before you even ask. In a sales negotiation, you'll learn how best to tailor your pitch to a customer's moods. All of which will make you more successful in reaching your own goals. The clues you glean can help you adapt the conversation to help convince the other person to think differently.

Next time you buy a car and are about to close the deal, ask the salesman whether that's really the best they can do. Then watch his or her facial response. I recently bought a new car, and during negotiations I happened to see the face of another salesperson sitting nearby. When I was told that I absolutely had the best offer available, the other salesman inadvertently communicated that perhaps more was available. He pressed his lips closer, literally signaling, *My lips are sealed.* When I quizzed my salesman further and asked directly whether they could improve the offer, his colleague subtly nodded yes. As a result, I got more for my trade-in, a couple of extra accessories for free, and a full tank of gas—all on top of what was supposedly the best deal possible.

Another example: Ask job applicants about what they want from a career with your company, and their facial expressions will tell you quite a lot. During interviews to fill a senior position within one of our research organizations, we asked what three separate candidates wanted from the role as well as where each hoped to be in five years. Verbally, they each assured us that they wanted to be team players, help us develop our business—all the usual pat answers. But facially they revealed a string of other emotions. First came contempt. (What an awful question!) Next was fear. (They suspected that we didn't believe their responses.) Finally, they revealed disgust. (We had detected their lies.)

In the end, none of the candidates got the job or even another interview. In time, we learned that all of them were in the process of setting up a competing organization of their own. They hoped to work for us only long enough to learn the ropes and pay their bills until they realized their true goals.

While every healthy person has an innate ability to read others' faces, not everyone realizes how powerful it is. To paraphrase Dr. Ekman, are you sure you want to know what others are thinking? Once you've got the power, you can't turn it off. Just as your newly discovered face-reading ability offers many advantages when it comes to communicating with others, so too do your own facial expressions say something to others about you. As you become more proficient at reading people's expressions, you'll also become better at managing your own.

It's extremely difficult to conceal your instinctive facial responses—but it's not impossible. Botox, for example, deadens the nerves that activate certain facial muscles . . . but that's a bit extreme for most of us. Some stroke victims lose the ability to communicate certain facial expressions, which inhibits both their verbal and physiological communication skills. Otherwise, our faces say so much about us—what we're thinking, what we're about to say, and even the type of life we've led. (Laugh lines speak volumes.) Sometimes, our faces betray feelings we wish they wouldn't, and we inadvertently send the wrong messages. Let's look at a few examples.

As we've discussed, a lot of women wear too much makeup. Ask for honest and expert advice on how to apply the right makeup in just the right amount. But maybe you don't use any makeup because your partner tells you how beautiful you are without it. How polite—and untruthful. According to several studies, men consistently find women in makeup more attractive than those without it. They judge made-up women as more attractive, confident, healthy, and even intelligent!

Shifting the perspective, women find the eyes to be one of the most visually attractive aspects of another person. Advertisers know that most women can't resist looking at babies' eyes, for example. (Remember the Gerber Baby? The child's eyes are part of what makes their branding so effective.) So, men, the next time you post a photo to a social-media or online-dating profile, drop the sunglasses. Use them in strong direct sunlight of course, but not because you think they make you look cool. They don't! Part of our success depends on how well we use our physical attributes. When others look at us, they want to see our eyes. When we want to communicate effectively, we do so in person where we can show our eyes.

Still, most of us have a lot to learn when it comes to looking our best to others. Understanding that can offer you a very powerful advantage. Spend a little time identifying and creating your ideal looks for attractiveness, dominance, or whatever message you want to convey. When you want to look intelligent, wear glasses. When you want to appear powerful and authoritative, wear a jacket and tie. A red tie in particular conveys strength and vitality. When you need to appear approachable and non-threatening, try knitwear. You'll gain a distinct advantage over others with whom you may be competing for a partner, job, loan—or anything.

One of our most powerful nonverbal communication tools is our smile, and research has shown that we have developed an advanced means of detecting what type of smile we're beholding. Put on an affected or staged smile to mask your fear or anxiety, and others will know it. Alternatively, if you exhibit a genuine smile, others will know that as well and feel a closer bond. But context is king. Revealing a genuine smile as you glance at your hand in a poker game may tip off your opponents. It's important to realize that you smile in different ways and you can use your smile as a positive part of your communications arsenal instead or using it as a barrier between you and others.

Now, how should you tilt your head? Never thought about it before? Now's a good time to factor it into your developing persona. Women should tilt their heads forward (chins down) to appear more attractive, and men should tilt their heads back to look more masculine. Combining the right sort of smile with the most appropriate head tilt will have a significant impact on how others regard you. Armed with that knowledge, you can manage how different people view you emotionally.

REMEMBER . . .

* Faces accurately communicate your real feelings and help you communicate more effectively.
* Facial expressions rarely lie, and it's nearly impossible to conceal instinctive facial responses.
* Makeup works because the coloration of people's faces send sexual signals.
* Women find the eyes to be one of the most visually attractive aspects of other people.
* Use your ability to read the faces of others to your advantage.
* Don't hide behind a polite smile.
* Use the right kind of head tilt.

Understand Personalities

W E humans are wired to get along with other humans, but not all
people get along with others. One of the core reasons that we like
some people and dislike others involves our personalities.

An individual's personality comprises behavioral, emotional, mental,
and temperamental attributes. Personality is predominately a psychological
trait, and we know that biological needs and chemical processes within our
bodies also influence it. We exhibit our personalities in how we behave and
respond within our environments, but it's not a one-way street. Our personalities in turn significantly influence our thoughts, beliefs, social interactions, and relationships.

Understanding your personality and the personalities of others offers an
extremely effective way to improve your life as well as others' lives. You may
want to increase how appealing you are to others, or you may want to excel
in your chosen profession. Whatever the goal, understanding what aspects
of personality influence different relationships—and then having the power
to align your personality more closely with those of positive role models,
superiors, and peers—gives you a good way to achieve more.

Academics and psychologists have identified five basic dimensions of
personality, often called the Big Five. In the 1970s, two personality-research
teams working independently developed a test known, unsurprisingly, as
the Big Five Personality Test. Paul Costa and Robert McCrae at the Natural
Institute of Health in Bethesda, Maryland, headed one group, and Warren
Norman of the University of Michigan's psychology department and Lewis
Goldberg at the Oregon Research Institute led the other. The two teams
had slightly different approaches to the study of personality but ended up
with the same results. They both found that most personality traits can be
reduced to one of five dimensions. The constituents of the Big Five are:

Openness to experience

Conscientiousness

Extroversion

Agreeableness

Neuroticism

These create the handy acronym OCEAN. Interestingly these five components appear attributable to genetics more than to a person's environment. In other words, your lineage influences your personality more than your surroundings. During young adulthood, a person's rating within each of the five factors may change. For example, average levels of agreeableness and conscientiousness typically increase, and extraversion, neuroticism, and openness generally decrease. Researchers have found, though, that after the age of thirty, stability prevails. That's not to say that your personality can't change given extraordinary circumstances or efforts to change them. It does indicate, however, that after age thirty people's personalities generally don't change *much*.

———

But several decades before the two research teams identified the OCEAN traits, psychologist Abraham Maslow laid important groundwork for our understanding of personality in the greater scheme of what makes us tick.

Best known for his hierarchy of needs—a theory of psychological health predicated on the need to fulfill innate human requirements before self-actualization can occur—Maslow stressed the importance of focusing on the positive qualities in people as opposed to treating them as a "bag of symptoms." He believed that we're all hardwired to go beyond what we are today to become everything we are capable of becoming.

Although Maslow was on the right track, our needs are somewhat more complicated than he detailed in his 1943 paper, "A Theory of Human Emotion." The hierarchy, he argued, begins with the basic physiological needs of breathing, water, food, excretion, sleep, sex, and homeostasis (the various internal systems that maintain our physical stability). In other words, what drives us most is basic survival. The oldest parts of our brains are programmed to seek whatever keeps us alive. Most of this programming is completely automatic, and it governs what we put into our bodies and what comes out. Our next fundamental needs are safety and security, employment, resources, morality, family, health, and property. Once we cover the physiological side of survival, then we focus on living our lives safely. That leads to the third layer, which is socialization: aligning with our social group, family, and sexual partners. Next comes esteem, which includes self-esteem, confidence, achievement, respecting others, and respect from others. Esteem manifests itself in our wanting to be higher up the ladder in our social group so that others will respect us. The fifth and final need is self-actualization, which includes morality, creativity, spontaneity, problem solving, lack of prejudice, and acceptance of facts.

Critics, readers, and other scientists considered Maslow's work profound at the time. The publication of his book *Motivation and Psychology* resulted in the creation of the field of humanistic psychology and pushed the scientific view of human nature into a more positive light, in which we're motivated to realize our fullest potential. Which is just how Maslow himself put it. He defines self-actualization as "the desire for self-fulfillment, to become actualized in what a person can potentially achieve." It's the desire to become more than we are and everything we can be.

When the five needs are illustrated in the form of a pyramid, the lower four contain what Maslow called the deficiency needs. They include esteem, love and friendship, security, and our physical needs. When these (except for the physiological needs) aren't met our bodies continue to function, although we feel anxious and tense. Then, at the top, sits self-actualization or the desire for self-fulfillment. That, after all, is what success is all about. Although Maslow was on the right track with his pyramid hierarchy, it lacks sufficient evolutionary input to be entirely relevant today. It's interesting to think about how meeting or failing to meet our human needs might alter our personalities, but we need to look at other indicators that contribute to who we are.

We humans are remarkably similar to one another. Each of has 23 chromosomes, more than 200 bones, more than 600 muscles, and we can expect to take 600 million breaths and have more than 2.5 billion heartbeats over the average lifespan. But as psychologist William James points out, "There is very little difference between one man and another, but what there is, is very important."

Over the last century, many psychologists have identified six different aspects of human behavior—called the Central Six—that establish the key differences among human minds. These six traits can be measured, and each person has a different combination of them. Social researcher Geoffrey Miller, associate professor of psychology at the University of New Mexico, refers to them as fitness indicators and itemizes them as general intelligence, openness to experience, conscientiousness, extraversion, agreeability, and emotional stability.

These core indicators provide an excellent means of linking Maslow's hierarchy with modern-day consumer behavior, which clearly reveals a great deal of the psychology behind personalities. On the one hand, we are wired to find the best mates possible. For men, that means finding women who will produce a large number of healthy offspring. For women, that means finding men who will help produce the fittest children and stick around to help protect them effectively. As we'll see as we look at the fitness indicators more closely, the aim behind nearly every purchase we make is to meet the requirements of one or more of the indicators, which offers a considerable leap forward over the rationale behind Maslow's hierarchy.

General intelligence is a relatively good index to genetic quality. According to Miller, general intelligence correlates positively with overall brain size, speed of performing motor tasks, physical health, mental health, and sperm quality. That, in turn, reflects well on general romantic attractiveness. Now think of all the intellect-related products on the market: brain foods, energy drinks, smart foods, and even smart drugs. Not even children can escape the onslaught, as many kids' games and toys are marketed as good for constructive play or brain training.

What goes against the logic of buying intelligence-boosting products is that we often pay more for status symbols that tell others at a glance just how intelligent we are. Examples include a barometer, a Rubik's Cube, and a pilot's watch. The last in particular costs serious money and only its owner can really appreciate it. But if a person believes the watch communicates his intelligence, then it's a sure sale.

Badges of intelligence like these look like an expensive waste of money—but that isn't quite true. Buying a status symbol can prove well worth the investment if only for the added confidence that it gives its owner. That pricey pilot's watch might give the man wearing it the self-assurance he needs to ask more women out on dates. More dates offer a greater chance of a sexual encounter, and the more sexual encounters, the greater his chances of populating the next generation with watch-loving, consumer narcissists.

The second fitness indicators is extraversion, the degree to which a person is friendly, outgoing, socially confident, and talkative. Extraverts enjoy leading, prefer being active, and exhibit high levels of self-confidence. People low on extraversion prefer working alone, tend to be physically laid-back and passive, and are less trusting. Low extraversion usually indicates both negativity and shyness.

As a fitness indicator, extraversion allows us to stand out from the crowd, to make a statement. Men often go to considerable lengths in the pursuit of extraversion: performing elaborate dance routines at nightclubs, commanding the center of attention at parties, and customizing their cars with special paint jobs, jacked-up suspensions, and deafeningly loud sound systems. All of those are ways of appearing more extraverted.

Women are more extraverted in social situations than they once were. This may have evolutionary roots since women historically tend toward group activity. Men often consider other men as potential threats or challenges, but women tend to share the caring for children with other females and to work as more of a team within the social group. Extraversion allows men to achieve more sexual encounters and women to find better quality mates—quantity versus quality. In both cases, from an evolutionary perspective, it's the principle of needing to stand out to have a chance of being selected that lies behind acts of extraversion.

Openness is the third fitness indicator. It relates to uniqueness, curiosity, and broadmindedness. Depending on our own degree of openness, we perceive others who are less open as dull, tedious, or boring. Conversely, we often regard those who appear more open than ourselves as a threat—bizarre, eccentric, or even downright mad. Although many of us try to appear more open than we really are, most of us have a natural comfort level of openness, and we seek others of a similar nature—and for good reason. From an evolutionary point of view, we have to be open enough to become part of a social group, but we also must fit into that group without being disruptive or burdensome to it. Miller cites evidence from a number of studies that indicates a lack of openness and some level of xenophobia may well have an evolutionary basis. But we all need some degree of openness to interact with others whom we can call to aid us in our quest for achievement.

The fourth indicator is conscientiousness. This personality trait includes characteristics such as punctuality, reliability, integrity, and trustworthiness. Conscientiousness is essentially the self-control exerted by the neocortex part of the brain onto the much more impulsive, selfish, short-term instincts of the limbic system. Conscientiousness matures slowly, but when it does, for example, it inhibits the impulsive mating activity that maximizes reproductive success among younger men. Because the neocortex manages conscientiousness, some researchers argue that this trait wasn't needed or even prevalent in our prehistoric hunter-gatherer lives. Therefore it's a more recent development.

Nevertheless, conscientiousness is important to everyone. We spend much of our adult lives trying to portray to others just how conscientious we are. We regularly wash our cars, keep our desks and front yards tidy, appear in public with ironed clothes and polished shoes. That leads us into the personal-grooming market, which, according to Miller, is misunderstood as a range of products for maximizing our physical attractiveness. But many personal grooming and beauty products actually serve as conscientiousness indicators. The clean-shaven man, the immaculately made-up woman, and the impeccably dressed business professional all communicate a strong sense of conscientiousness. Since conscientiousness goes hand in hand with hard work and success, scientists argue that those of us who exhibit the most conscientiousness will make the best mates. The men will provide better for the family while the women will take better care of the young.

Miller calls the fifth fitness indicator agreeability, which lies at the heart of altruism and progressivism. It's that rare product of natural selection that allows us to rise above the instincts of most of the rest of the animal kingdom. We can see our agreeable nature in many aspects of modern life. Driving a gas-guzzling SUV exhibits a minimum amount of agreeability, while driving a hybrid compact sedan communicates significantly more.

Agreeable people try to get along with as many people as possible, so they tend to conform. Conversely, those who exhibit low agreeability tend to be dominant, assertive, and iconoclastic. To test that premise, Vladas Griskevicius and his team at the Carlson School of Management at the University of Minnesota conducted a study that compared men exposed to sexually arousing images with men who weren't in any way sexually primed. The primed men showed less conformity and more dominance than the others. Women have a stronger preference for a mate who displays dominance, assertiveness, and a degree of risk taking—all viewed as admirable traits in a male. A follow-up study by Griskevicius and Miller reported that whether a man is evaluating subjective preferences or hard facts influences the effects of sexual priming. Sexually primed males show strong nonconformity when making subjective choices as to which make of car they prefer, but they show strong conformity when responding to fact-based questions of general knowledge.

Women, on the other hand, show stronger conformity when sexually primed but remain neither agreeable nor assertive when responding to factual questions. According to the findings of the study, men want to show off their bad-boy nature when they want to impress women, while women want to show off their agreeability when they want to impress men. Agreeability is a positive trait overall, but in the competitive world in which we live sometimes we must put our own needs and desires before what others think of us.

The sixth and final indicator is emotional stability. It measures the amount of resilience we have, how resistant we are to stress, how quickly we can recover mentally from an emotional setback. The emotionally stable tend to be calm, relaxed, optimistic, and quick to come to terms with a given situation. Those low in emotional stability are more likely to exhibit anxiousness, depression, and pessimism. They are quick to become angry, tend to cry with little or no provocation, and suffer from anxiety.

Emotional stability offers both genders a more solid, reliable, dependable partner. We have a strong desire to project ourselves as emotionally stable both from a mating perspective and from the point of view of social acceptance. Within the workplace in particular, emotional stability serves as a key fitness indicator. You wouldn't want a neurotic, distraught person managing air traffic control at JFK, nor would you want a cold, unemotional person in the PR department.

So what began in the early 1940s as Maslow's hierarchy of needs has evolved into Miller's fitness indicators, which give a more reasoned list of needs that incorporates modern scientific knowledge with a solid understanding of human evolution.

Miller argues that the root of consumer capitalism—our need to acquire things—lies in biology, and my own research proves that's true . . . to a point. As humans evolved in small social groups, both image and status proved critical to social hierarchy and the ability to attract suitable mates. From that perspective, the fitness indicators offer a solid and generally accepted argument as to why we think and behave as we do. Miller also believes that consumer capitalism is largely an exercise in gilding the lily. He holds that humans have wondrously adaptive capacities for intelligence, kindness, creativity, and beauty and then forget how to use them for making friends and attracting mates. Because of that shortcoming, humans rely on things—that is, goods and services—to advertise their personal traits to one another.

While that's true, again to a point, it's irrelevant whether such purchases directly alter our status within social groups or merely give the perception of status. When we buy status badges, we do so in the firm belief that the ownership of them alters others' perceptions of us. Whether people's perceptions change in the way we believe they do remains irrelevant. If by making the purchase we believe that we're altering others' perceptions, then the purchase itself becomes a valid fitness indicator. That's what we mean by dressing for success. The man standing at the bar in his brand new designer jeans chatting to a young woman may believe on some level that the quality of his pants impresses her, but even if it doesn't, his belief arms him with the confidence to talk to her with the aim of dating and mating. The investment in the clothing has had a positive impact on the purchaser's goal.

To help identify your own personality traits and to recognize the inner motivations that drive you, here's a short test.

Exercise
Personality Traits

Use the rating scale below to determine how accurately each statement describes you. Your answers should reflect who you are, right here and now, not as you wish to be in the future. Describe yourself as you honestly see yourself in relation to other people you know of the same gender and age. Read each statement carefully, and then circle the number that corresponds most accurately to the numbers on this scale:

1 Very Inaccurate

2 Somewhat Inaccurate

3 Neither Accurate nor Inaccurate

4 Somewhat Accurate

5 Very Accurate

GENERAL INTELLIGENCE

I prefer deep meaningful conversations to idle chat.

1	2	3	4	5

I feel proud when asked for my advice or opinion.

1	2	3	4	5

I don't like to keep my thoughts to myself.

1	2	3	4	5

I frequently voice my opinion in discussions.

1	2	3	4	5

AGREEABILITY

I sympathize with others' feelings.

1	2	3	4	5

I'm interested in other people's problems.

1	2	3	4	5

I'm a team player.

1	2	3	4	5

I feel others' emotions.

1	2	3	4	5

CONSCIENTIOUSNESS

I don't often make a mess of things.

1	2	3	4	5

I get chores done right away.

1	2	3	4	5

I often remember to put things back in their proper place.

| 1 | 2 | 3 | 4 | 5 |

I like order in my life.

| 1 | 2 | 3 | 4 | 5 |

EXTRAVERSION

I'm the life of the party.

| 1 | 2 | 3 | 4 | 5 |

I talk a lot.

| 1 | 2 | 3 | 4 | 5 |

I don't like to stay in the background.

| 1 | 2 | 3 | 4 | 5 |

I talk to a lot of different people at parties.

| 1 | 2 | 3 | 4 | 5 |

OPENNESS

I can easily understand unrealistic ideas.

| 1 | 2 | 3 | 4 | 5 |

I have a vivid imagination.

| 1 | 2 | 3 | 4 | 5 |

I prefer intangible ideas to practical solutions.

| 1 | 2 | 3 | 4 | 5 |

I am highly creative.

| 1 | 2 | 3 | 4 | 5 |

NEUROTICISM

I'm stressed most of the time.

| 1 | 2 | 3 | 4 | 5 |

I often feel sad.

| 1 | 2 | 3 | 4 | 5 |

I get upset easily.

| 1 | 2 | 3 | 4 | 5 |

I have frequent mood swings.

| 1 | 2 | 3 | 4 | 5 |

Here's how to score your answers: For each of the indicators, add up the numbers you circled for each section (General Intelligence, Agreeability, Conscientiousness, Extraversion, Openness, Neuroticism). To see how you compare with others, write your scores below, next to the average scores on the test.

General Intelligence	10	
Agreeability	8.4	
Conscientiousness	9.8	
Extraversion	10.8	
Openness	11.4	
Neuroticism	9.6	

Interestingly, men and women show differences in their Central Six scores across cultures, with women scoring higher in both agreeability and neuroticism and men scoring the same or higher in the other four domains. All of which raises the question, how are the brains of men and women the same or different? Let's take a look in the next chapter.

===

REMEMBER . . .

* You take your personality with you wherever you go.
* Men show off their rebellious nature when they want to impress women.
* Women show off their agreeability when they want to impress men.
* Understanding your personality helps you deal with life's daily challenges and achieve more of what you want.

{32}

Mars and Venus

WHEN it comes to normal, healthy brains, there's not much difference between them. Men, women, Asians, Africans—it really doesn't matter because we're all pretty much the same. It's a mundane truth, but differences do exist, even if only slight, and they play a marked role in how you can approach achievement. Some of them might even surprise you.

Let's get straight to it. In our ancestral societies women cared for the family group while the men went off hunting. As a result, women developed better and subtler interpersonal communication skills than men. By nature, women's brains are more group-oriented and apt to seek solutions by discussion. Women also tend to have better communication skills and emotional intelligence than men. Men can have trouble picking up on emotional cues unless they're spoken clearly or in-your-face obvious, a limitation that often makes for tricky communications between the sexes.

Ladies, if you want to communicate effectively with men, forget the subtlety. Most men simply won't get it, and you'll think them insensitive and boorish. Lay it on the line, and you'll fare much better. Men, to communicate effectively with women, listen more closely. Research has identified that women like to share their problems, but they're not necessarily looking for an immediately viable solution so much as acting out that old maxim that a problem shared is a problem halved. Women have evolved to value empathy, so learn how it works, men. If you're still in the dark, put yourself in her position emotionally. Empathy feels almost as strong as if you yourself were experiencing the same emotion, tying back to our earlier discussion of how mirror neurons work in the brain.

We all experience stress at one time or another. How we handle it depends at least partly on evolution. When faced with stress, women use more of a group-nurturing response. In many mammals, including humans, females form tightly knit, stable attachments with other females. This tendency increases in times of duress. Among modern women, you can see this habit in action in hours-long phone calls, book club meetings, girls nights in, and long coffee or wine-bar dates. According to Shelley Taylor, a professor of psychology at UCLA, befriending is "the primary gender difference in adult human behavioral responses to stress."

Part of the differences in how we cope with stress has to do with women having larger limbic systems, the part of the brain responsible for our emotional life and that is the key to the formation of memories. Women are more in touch with their feelings and better suited to expressing their emotions. This makes them better at connecting with others but, unfortunately, more prone to negative emotions and even depression.

Men, on the other hand, usually employ the fight-or-flight strategy by preparing to fight what the brain has decided the real problem is *and* what the root causes are. In other words, men analyze ulterior motives and play the Why, Why, Why? game to try to solve the problem. Women often find this behavior frustrating, because by employing that problem solving strategy men fail to offer emotional comfort in times of need. Women, you need to spell it out one word at a time to make it clear. If you want him to listen and not to solve the problem for you, you really have to say exactly those words to him. If you know a man exhibiting a lot of stress, try to help him by suggesting that he refocus on his problems. Encourage him to think of how his anxiety is impacting the people whom he cares about. Nudge him away from fight-or-flight syndrome and toward tend-and-befriend.

Both men and women can improve how they deal with stress by realizing that the sexes have different coping strategies. Instead of getting frustrated with the opposite sex, encourage him or her to see the situation from your point of view. Male and female brains differ in other ways, but the point to remember is that we're more similar than different. Understanding the differences can help your personal relationships as well as your business relationships, particularly in dealing with members of the opposite sex.

REMEMBER . . .

* Women tend to have better communication skills and emotional intelligence, and they cope with stress by forming nurturing social groups.
* Men cope with stress by employing the fight-or-flight strategy, analyzing ulterior motives, and trying to solve a problem.
* To communicate effectively with men, lay it on the line, clearly and simply.
* To communicate effectively with women, listen closely and empathize.

{33}

Dual Coding

WE'VE established that communication between people plays a key role in how attractive they'll find one another. This may manifest as physical attraction, or it may surface as a simple preference, one sales pitch preferred over another or one teacher more popular than the rest. Now let's explore how to communicate so that others process more effectively what you're saying or implying.

Efficient and effective communication means having your audience remember the key aspects of what you tell them. Successful communication depends on how much of the content the recipients absorb and internalize. That's especially true in complex areas such as education or business, where several suppliers may find themselves vying for a single sale.

Earlier we learned how humans have both long-term and short-term memories. Strategies to aid recall that involve simultaneous visual and verbal processing—known as dual coding—are significantly more powerful than either verbal or visual coding alone. Dual coding helps information pass from short-term memory into long-term by creating multiple retrieval routes for the same piece of emotional information.

Allan Paivio, professor of psychology at the University of Western Ontario, and Kalman Csapo of London Psychiatric Hospital recognized the long-term benefits of dual coding by identifying that concepts presented once simultaneously in visual and verbal form were remembered better than the same concepts presented twice in just one of those forms. When you communicate using dual coding, you're involving more than one sense to convey the meaning of your message. Here are some examples.

As a teacher, you can talk to your class about a particular subject, but more of your meaning will enter the students' long-term memories if at the same time you show images associated with your subject matter. By using this process, your students have to use different processing parts of their brains at the same time—in this instance, both auditory and visual processors.

Want to make an impression at that next big job interview? Use dual coding to do just that. If you decreased spending and increased profits at your last job, as you're saying so produce an image of the data. Give your interviewer a bold graph on a sheet of paper that she or he can keep. Do you want to stress that you have extensive contacts in a particular field? Hand the interviewer a printed list of all those names. The interviewer will remember more about you and what you had to say.

Perhaps you're a car salesman telling a customer what a comfortable ride a particular BMW has. She'll get a stronger perception of that statement if she's sitting in the luxurious leather seats of one of the showroom models. Focus on the tactile experience—how the steering wheel feels, how the seats feel.

Another example: Suppose you want to tell your young children something they need to remember. Relying on dual coding, you tell them to look both ways before crossing the road as you yourself look to the left and then right.

Whatever the situation, always remember that simultaneity is key. Dual coding is most effective when the brains of those receiving the communication must process information from different senses *at the same time*.

In the next chapter, let's learn more about making good impressions and how you can optimize the power of your communication by utilizing not only words but also tone of voice, inflection, and nonverbal communication.

REMEMBER . . .

* Simultaneous multi-sensory communication is more memorable than single-sense content.
* Combine sensory data—aural, tactile, visual—to make more of an impression and impact.

The 7-38-55 Rule

ARLIER we saw that words represent only 7 percent of human communication related to feelings and attitudes, tone of voice accounts for 38 percent, and the remaining 55 percent comes from nonverbal sources, such as body language and facial expressions. Think about that for a second. The English language has over one million words, and we've created complex rules to pronounce and use them, but we *barely* use the spoken word when communicating about our emotions.

People often misconstrue the 7-38-55 rule because they overlook the vital phrase "related to feelings and attitudes." An example: If I gave you directions on how to get from point A to point B, I'd use significantly more spoken words than nonverbal communication to make myself clear, but if I tried to explain why point A made me unhappy and point B made me glad, you'd find more of my meaning in how I sounded and how I behaved than just the words I uttered. Understanding this rule has significant ramifications when striving for greater personal achievement because it impacts not only how you communicate with others but also how you understand yourself.

Text messages (excluding emoticons and emoji), e-mails, memos, and letters are limited forms of communication because they convey only words. Unless you're a *very* good writer, intended tone or emphasis is lacking. Facial expressions are similarly absent. Talking on the phone can convey words and the way they're said, but it can't incorporate the messaging revealed in facial expressions, such as rolling your eyes or smirking. Obviously, these methods are convenient—otherwise we wouldn't use them—but they do leave something to be desired.

We're also pretty poor practitioners of English when it comes to describing things. If I asked you to point out the face of a friend in a group picture, you could do so in a fraction of a second. But if I asked you to describe the elements of that friend's face so I could pick her out, sight unseen, you'd have difficulty completing the task. In fact, unless she had a distinguishing visible feature such as a birthmark, facial piercing, or purple hair, you might not be able to complete the task at all.

Videophones and teleconferencing are great new forms of communication, but they, too, fall short as mediums of communication because facial expressions can change in a heartbeat, and video typically consists of no more than twenty-five frames, or still images, per second. Video misses many of the expressions on a person's face because minute changes of expression often fall between those frames. Also, because of the data compression used, there's usually a lag between the video and audio segments, which can cause emotional dissonance between what you're seeing and what you're hearing.

So when you want to have a heartfelt discussion with someone, do it in person. Face-to-face communication can be intimidating, but emotions usually are more persuasive than cognitive reason. If you want to communicate your feelings along with your words—which in important situations you do—you'll need to do so in person.

Consider these two alternatives. Imagine that your partner or parent sends you a text message that reads "luv u." Now imagine if he or she looked you in the eye, held your hand, and said, "I love you." The two sentences convey the same basic message, but they have vastly different emotional impacts.

Salespeople know it's easy to send out millions of letters and e-mails to potential customers, but that's what's called a *spray-and-pray* technique. Spending the same amount of time and effort on developing ways to get in front of the ideal customer is a much better way to undertake outreach. If you work in an industry that has regular conferences or trade shows, visit the exhibitors' booths to rub shoulders with other visitors, perhaps share a coffee, and ask attendees if they're available for a meeting later. That approach has a far greater emotional impact than any other means of communication.

Communicating face-to-face creates a much more personable and engaging encounter. Each party can pick up on the mood of the other and figure out what each person's attitude toward the conversation is before deciding whether to delve into deeper detail, change the topic, or terminate the discussion entirely. Even if major cultural differences exist between individuals—such as an extreme age gap or a language barrier—the parties can still have a fruitful meeting because face-to-face interactions are universal. A smile is a smile in every age range and every language. Eye contact, facial expressions, and body language all offer additional means of communicating that you can't get in an e-mail or letter.

In short, if you have something meaningful to say, something with feeling or attitude, do it in person if you can and remember the 7-38-55 Rule.

REMEMBER . . .

* Go for face-to-face communication when you want to convey emotions.

* Emotions are more persuasive than cognitive reason, so persuade in person.

* Face-to-face interactions provide the advantage of not being restricted to only words.

{ 35 }

Mirroring

BACK in the 1970s, psychologist Richard Bandler and linguist John Grinder created a relatively modern approach to communication, personal development, and psychotherapy. It's known as neuro-linguistic programming (NLP). At the time it was, as the saying goes, all the rage. But more recently it has been partially discredited, and the approach has been characterized as pseudoscientific. Polish psychologist and writer Tomasz Witkowski wrote in 2010 that "NLP represents pseudoscientific rubbish, which should be mothballed forever."

But some aspects of it are scientifically proven to be effective. The trouble is that critics tossed out both the bad and good components of NLP when they should have been separating its individual strengths and weaknesses. Let's look at the science behind why mirroring and representational systems work and then make them work for you.

Mirroring, in short, is copying a person while interacting with him or her. The behavior may include mimicking gestures, movements, rhythms, body language, breathing and other aspects of communication, and you can often spot such behavior between couples and close friends. Mirroring is also a very useful communication tool. It's easy to be drawn to people who are good at mirroring. They are announcing, in effect, that they are like you. Even so, vastly different abilities to mirror others exist, and as with most skills, people tend to get better with practice. The key to effective mirroring is subtlety. When done right, good mirroring is perceived subconsciously.

The easiest forms of physical mirroring for you to learn and harness are body language and breathing. If a man sitting opposite you has crossed his legs, you can mirror him by doing the same. If he folds his arms, do likewise—but not simultaneously! Remember, mirroring isn't a game of mimicry. When the process works, it works because if other people perceive that you're alike, the less they'll fear you and the more they'll have an affinity for you.

According to Piotr Winkielman and Liam Kavanagh of the psychology department at the University of California at San Diego, "Mirroring is an important part of social intelligence as well as interaction." It follows, then, that it's equally important to know when not to mirror. The success of mirroring depends on mirroring the right people at the right time and for the right reasons. Sometimes the most prudent course of action is *not* to imitate someone.

Winkielman concludes that "It's good to have the capacity to mimic, but an important part of social intelligence is knowing how to deploy this capacity in a selective, intelligent, context-dependent manner, and understanding, even implicitly, when mirroring can reflect badly on you." For example, do you want that guy over there to think that you like him? If you're in a one-on-one conversation with another person, mirroring can help develop a rapport between you. But if others are involved, such as in a group setting, they may judge your behavior based on the person you're mirroring and then form faulty perceptions of you. If you mirror a less than competent interviewer and others see you, they might perceive you as incompetent, too.

Of course, not all mirroring is or needs to be physical. You can also mirror words. British professor Richard Wiseman at the University of Hertfordshire conducted a study in which food servers mirrored their customers by repeating their orders back to them, while others responded with positive affirmations such as "great," "good choice," and so on. The results showed the influence that mirroring can have. The servers who used the mirroring technique received on average a 70 percent larger tip than those who used only positive affirmations.

But as with gestures, when it comes to using words as forms of mirroring, don't copy the other person directly. Be subtle. Listen for patterns of language or particular but common phrases and then drop them into your own conversation.

Representational systems are the inner processes each of us uses to describe things and events from a sensory perspective. For example, Einstein credited his discovery of the theory of special relativity to a mental visualization strategy of "sitting on the end of a ray of light," and many people create visualizations in their minds as part of their decision-making processes.

Back in the stone ages of NLP, researchers believed that individuals had a preferred representational system based on a visual, auditory, or kinesthetic thought pattern. Visual people used words such as *look, see, picture, illustrate, observe, notice,* and *imagine.* The auditory crowd favored words such as *hear, listen, rings a bell, sounds like,* and *I hear what you're saying.* Those with a kinesthetic preference spoke with words such as *touch, feel, handle, grip, light, heavy, grasp,* and *comfortable.*

Critics discredited this part of NLP by the early 1980s, because scientists agree that most people use all of their senses—whether consciously or unconsciously. While one system may seem to dominate, that dominance often depends on context. In other words, it's not that Bob or Emma is a predominantly visual person, but that the context, such as talking about a trip to the movies, favors a visual experience. So it wasn't that all of NLP was wrong but rather, at least in the case of representational systems, its interpretation of how our brains use them was in error.

Pay attention the next time you talk to someone. If a woman tells you that she danced for hours to a live band before striking up an amusing conversation with the bartender, that's primarily an auditory experience. Once you understand how she represents a particular situation mentally, you can build a solid rapport with her by adopting the same representational system when talking about the same subject. Add some auditory references to your side of the conversation, and notice how quickly she aligns herself with you emotionally.

Now that we understand what drives attraction and, in particular, how we communicate with others and how they communicate with us, in the next chapter let's take a quick look at a couple of ways of handling difficult social situations.

REMEMBER . . .

* Mirroring draws people emotionally closer.
* The key to effective mirroring is subtlety.
* Know when *not* to employ mirroring as a rapport strategy.
* Identify the sensory words a person uses in conversation and use those words yourself.

Cope with Social Anxiety

SOCIAL anxiety is an intense fear in group situations that causes considerable distress and impaired ability to function normally in parts of
everyday life. Generally, social anxiety involves a persistent, chronic fear
of being judged by others and of being embarrassed or humiliated by your
own actions. Perceived or actual scrutiny from others can trigger these fears.

People who have it may recognize that the fear of social interaction
may be excessive or unreasonable, but overcoming it can often prove quite
difficult. Physical symptoms include blushing, sweating, trembling, racing
heart, nausea, and stammering. In just one example, it might manifest itself
in an employee making a presentation to his superiors. If he hesitates or
stumbles over a particular word, he may fear that his bosses caught the
mistake and now think less of him as a result. That thought process easily
leads to further anxiety, resulting in additional stuttering, sweating, and,
potentially, a presentation disaster.

According to psychologist B. F. Skinner, sufferers typically address these
types of phobias by enacting escape and avoidance behaviors. For example,
the employee above might excuse himself from the room in a hurry before
the presentation is done (escape) and refrain from giving future presentations (avoidance). Recognizing and preventing these automatic responses is
a good way to address any social anxiety you may have. Remember, your
brain is wired for you to be social, so it's better to work with it.

Fortunately, if you do find yourself suffering from social anxiety, you can
employ several techniques to address the problem. One of the most effective
is the Why, Why, Why? game. In it, you think through the situation and
break down your anxiety into more manageable chunks. If you fear making
a presentation to your boss, ask yourself why. Keep asking yourself this same
question until you uncover the real root cause of your fear.

Overcoming Presentation Jitters

One of my colleagues was overwhelmed at the thought of delivering a keynote presentation to the senior management of a global cosmetics company. He knew his content well and was an expert on the subject, but he became fixated on the possibility that he might screw up. Not surprisingly, he made a mess of the entire presentation. The result was an angry senior-management team and an uncomfortable situation for everyone involved.

I helped my colleague dissect his problem, and he identified the root of his trepidation as a fear of not being able to handle a question from the audience. He was mortified at the thought of not knowing the answer to a question that hadn't even been asked! I advised him on how to confront his fear for his next corporate presentation.

A couple of weeks later, he stood in front of another team of executives. Before beginning, he announced that, while he was a respected authority on the subject matter, the presentation he was about to deliver was created by a team of experts whom he might have to consult to get the best answer to any questions the audience might have. He also asked that they hold their questions until the end of the presentation, thus freeing him from worrying about them as he spoke.

He gave a wonderfully engaging presentation, professionally delivered and well received. In fact, the cosmetics company asked him to present to other departments in their organization on a number of occasions since then. He delivered the same presentation to head-office teams in a number of different countries, but the best part was that he genuinely looked forward to each presentation he gave.

If you believe you have some form of social anxiety, candidly explore what lies behind it. Why don't you like talking to strangers? Why are you afraid of what they might say? Why do you worry that they'll put you down? Why do their opinions matter to you? More often than not, the underlying cause will turn out to be something much less troubling and insurmountable than what your brain had created.

REMEMBER . . .

* Your brain is preprogrammed for you to be social, so don't fight it.
* If you do have social anxieties, use the Why, Why, Why? game to overcome them and other negative emotions or impulses.

{37}

Deal with Naysayers

As we've seen, humans get along with one another by working in social groups in order to feed and protect one another. Since this is an ancient evolutionary trait, a lot of the aspects we seek in others are predominantly emotional. If we see a friend crying, we ask what's wrong. If someone we know is frightened, we try to put him at ease. That's because, as a species, we care about the feelings and emotions of others, particularly those in our immediate social groups, including family, friends, classmates, and coworkers.

Just as our own brains protect us from the potential pain of failure (which really are just alternative outcomes anyway), so too can the brains of others offer protection. But the problem with others trying to protect us is that they may steer us astray.

Let's assume that you've resolved to lose weight. To aid you in your goal, your brain prompts you to tell others in order to elicit their support. Unfortunately, too many other people automatically prevent us from attaining our goals by putting up a wide range of roadblocks. "What diet is it this time?" or "Yeah, I tried that diet, but all the weight came back as soon as I stopped." In another scenario, you may want to enhance your education and employment prospects by taking evening classes. Once again, you can expect a range of doubting-Thomas types to brandish a wide range of reasons not to enroll—it's exhausting, expensive, time consuming, and there are no guarantees that it will work.

The good news is that once you accept that your mind will try to dissuade you from achieving success to spare you the pain of failure—and recognize, too, that others may attempt to keep you from your goals—you can develop a strategy for more meaningful change. Here are some techniques for dealing with the negativity of others.

For starters, don't let the thought of someone naysaying deter you from sharing your goals socially. The very act of going public about your targets and aims strengthens your brain's determination to bring them to fruition. A good technique for disabling the negative influence of others is to broaden the focus of your success so that others sense an emotional benefit from it.

For example, if one of your goals is financial independence so that you can provide for your family, include the naysayers in your list of potential beneficiaries. Tell them you want to succeed so you can spend more quality time with them as a result. Or tell them that once you succeed, you're going to take them out to dinner to celebrate. Now you've shifted their focus away from how your goals might affect them negatively (making them look fatter or poorer or not seeing you as often) to something that benefits them emotionally.

As another example, let's say you're looking for a new romantic relationship. Because your friends are trying to protect you from the potential hurt of failing, they may disparage whomever you decide to pursue. They might also fear that your new relationship will weaken or supersede your existing friendship. Selfishly, their brains will try to optimize your existing bonds while dissuading you from looking for other relationships—especially lovers. Your friends rarely will have hidden agendas or outright malice toward you, but an evolutionary urge may be driving their urges to keep you firmly in their social circle.

Once you understand that others' motives may be subconscious and sometimes inappropriate, you can grasp their points of view more readily without acquiescing to them. Take their comments and reactions with a metaphorical pinch of salt because you understand what's driving them to think the way they do.

As a final point on the subject of the influence of others, which might seem obvious but is worth emphasizing, seek out other successful people and avoid the losers. Here's a good example. When an average high handicap golfer teams up with better players, his or her own game tends to improve. He or she concentrates harder and pays more attention to how the other players hit the ball and manage the course. I have seen this for myself many times and analyzed the golf scores of others to validate the phenomenon. In the company of upbeat and self-motivated people, your motivation will soar. The reason it works has to do with mirror neurons, as we discussed, as well as our evolutionary hardwiring to fit in with similar social beings—even if only to play a better game of golf!

The Problem Is . . .

I recently gave a lecture at several universities that entailed conducting joint psychological studies with both students and professors. At one university, the professors often began sentences with, "The problem is . . .", a phrase that did little for their creativity or mine. After several meetings, I retrained their communication methods so that whenever they even thought about starting a sentence with such a limiting phrase, they immediately corrected themselves and replaced it with, "The opportunity here is . . ." Since then, they have rarely expressed the blind negativity of the past.

That brings to mind another verbal tripwire to avoid. The word *but*, when used to counter a thought, nearly always acts to dampen a group's enthusiasm. "It sounds like a good idea, *but* . . ." or "That's fine in theory, *but* . . ." In both cases, the word acts as a killjoy by countering positive thoughts with negative ones.

How to work around negative conversations? Replace *no* or *but* with "Yes, and . . ." You may get a few quizzical looks, *and* you'll be amazed at how quickly the conversation shifts toward the positive!

REMEMBER . . .

* Don't let others protect you from success, and don't let naysayers deter you from sharing your goals with them.
* Going public about your goals strengthens your brain's determination.
* Without knowing it, your friends might be trying to hold you in their social circle.
* For people who express negative reactions to your goals, include a reward in those goals that benefits them.

PART FOUR

CREATE A MINDSET FOR SUCCESS

※

{38}

How Belief Systems Form

I N our brains, the prefrontal cortex relies on the different types of memory, which combine to create beliefs—the collective codes by which we live our lives. At its most obvious level, your belief system is the set of ideals that governs your thoughts, words, and actions.

We're born without any pre-existing knowledge of the society we're entering. To survive, we make observations and draw conclusions from them. Formulating a belief system is essential to our survival and perhaps even to consciousness itself. These beliefs can be cultural (it's not acceptable to go to the supermarket in the nude, for example) or practical (when hungry it's unwise to eat your own limbs). Beliefs are the simplest form of learned concepts and therefore among the integral components of conscious thought.

Typically, beliefs fall into three categories: core beliefs (those that the thinker may be actively considering), character beliefs (those that the thinker may never have considered before), and cultural beliefs (the standards and expectations that a culture has created and by which its members abide). Although we often think of beliefs as religious or political in nature, they needn't be. For example, if asked, "Do elephants wear bedroom slippers?" you would answer no, which would fall in line with your beliefs about the natural world, despite never having thought about the possibility before.

Psychologists who study beliefs suggest that they can form in a number of ways. In our early development, we tend to internalize the beliefs of the people around us. For example, many of us believe the tenets of the religion we learned in childhood. These beliefs may be adopted as core, character, cultural, or any combination of them. Later in life, we also adopt beliefs from charismatic leaders or celebrities, and we internalize these beliefs even if they contradict our previously held beliefs or produce actions not in our own best interest. (As Lincoln Barnett, editor of *Life* magazine, said of

Albert Einstein, "Common sense is actually nothing more than a deposit of prejudices laid down in the mind prior to the age of eighteen.")

For example, young adults sometimes develop the belief that they need to look like the stick-thin models on the covers of fashion magazines, which causes some of them to develop eating disorders. More positively, Princess Diana reduced people's aversion to contact with AIDS patients when she publically engaged with and touched them. On one occasion in Harlem, Princess Diana asked a seven-year-old suffering with the disease: "Are you very heavy?" She then bent down, picked him up, and hugged him for two or three minutes.

The belief of relationship by association with leaders or celebrities often interlinks with our roles as consumers. If you wear a football jersey with the number of a tough linebacker on it, you acquire some of that toughness. In this case, toughness translates into becoming an attractive, successful, above average mating candidate. Of course, a significant gap may separate the perception and the reality, although not necessarily in the mind of the shirt-wearer.

You can also form beliefs without absorbing them directly from outside sources. Some singers—whose parents tell them they have talent and who receive applause at karaoke bars—begin to believe in their own ability so much that some of them go and audition for shows like *American Idol* and *The X Factor*. For most, the judges' comments crash-land them back to reality in just a few seconds. In rarer cases, people can develop their mistaken beliefs internally because of a brain injury or dysfunction.

Belief systems can be headstrong and reckless because, like a lens, they distort both what and how we see. People have killed one another for millennia over religious and political ideologies. Then again, some belief systems—like the opposing examples of stoicism and hedonism—can help people endure and even enjoy life. The systems always appear sound and genuine to their believers even when confronted with empirical evidence to the contrary.

But the important point about belief systems is that what we learn in life we can unlearn—for better or worse. Those conscious of how their beliefs form can adopt new beliefs and act on them against their own interests, as we saw above. That means that celebrity endorsements and life experiences can change how we see the world. As we live our lives, we're constantly reconciling our perceptions of reality with our beliefs. In cases where some aspect of that perception isn't present or possible, we either change our beliefs or oppose the perception. Businesses recognize this behavior and target our belief systems accordingly. That happens most often in advertising, which has a number of guiding credos. One is that repetition forms beliefs. Another is that strong positive emotions form from associating beliefs with images of happiness, love, sex, and success. The next time you see an ad, think about how it's attempting to target your belief systems through the use of emotions. Here are a few memorable slogans: "Be yourself. Nonstop." "When you care enough to send the very best." "You've come a long way, baby!" These sayings illustrate the connection that admen try to make between emotions and beliefs.

The good news is that we can use our belief systems to rise above impulsive thoughts and automatic urges, allowing us to make more intelligent, achievement-driven decisions. To do that, we need to use as much of our brains as possible. The first filter through which we process information is emotion (again: fight, flight, fornicate). This step occurs so quickly that it's hard to alter it—though not impossible, as we've seen. That's why it's so important to align your goals emotionally. Next, cognition tries to determine the best course of action based on reason and our beliefs. Often this is where errors in judgment enter our lives, because the rational part of the brain is under constant pressure to make objective, reasoned decisions about absolutely everything (which is impossible) and also because our beliefs are often distorted.

If you go through life blaming yourself for your lack of success, for example, then in your mind that perception will become fact, and that fact will embed itself as a belief. In that situation, when you're making an instinctive decision, cognition feeds it into the distorted belief that you won't succeed. Your own cognition, in other words, is working against you. But this is also where self-motivation—psyching yourself up—and other communication tools can help. They can help rewire your negative belief by associating a new and more productive belief with a specific trigger that, over time, becomes more powerful than the principle it overrode. The process weakens the old belief, and the more positive alternative takes its place.

Here's an example. One of my business associates once worked as a sales representative. He had serious problems making appointments and closing deals, so much that he developed a fear of the telephone. He became so used to receiving rejections by phone that his brain assumed that whenever he picked up the phone, he would be turned down again. In time, he stopped using the phone altogether, making all sorts of excuses as to why the phone was an ineffective sales aid. He switched to communicating by only through e-mails and letters, and his productivity declined.

He had a real desire to sell, though, so he needed to realign his belief system regarding the telephone. He desired financial success because he wanted others to perceive him as a significant member of society. To realign his beliefs, we placed images of the material possessions he craved, including a $2,000 racing bike, around his telephone so that each time he reached for it he literally was reaching for success. That innocuous communication device that had enabled his debilitating belief suddenly became a conduit to his self-worth. After that small step, my colleague soon became the top-performing salesperson in the company.

In order to learn more about your personal belief systems, here's a simple exercise you can do.

Exercise

Analyzing Your Beliefs

First, identify a goal that you want to achieve but haven't. Write down all the reasons you've failed, playing the Why,

Why, Why? game. (Why haven't you remodeled the kitchen? Why can't you save enough money? Why do spend all the spare cash you have?) Continue until you're sure you've drilled down to the root cause of failing to achieve what you want.

Once you've done that, identify the negative belief that's holding you back. Now pinpoint when your failure became a belief. Was it something your parents or a teacher said long ago? Is your brain protecting you from something you've experienced before? By flushing out the root cause, you'll make it easier and more natural for your brain to take on a new positive belief. Then you can introduce the replacement belief by using positive self-motivation, emotionally empowering imagery, and whatever other devices link a specific trigger (such as a ringing telephone) into your preferred course of action.

Your belief system has been developing since birth, and over the years it has undoubtedly changed. Various events have fed and fuelled what you believe today. Sometimes those beliefs improve you and your life, but on other occasions they can damage your quest for success. If you really want to change, first you need to believe that change is achievable. That takes action on your part, so do it fast. The quicker you act, the sooner you can expect results. To quote Victor Kiam—entrepreneur and president of Remington Products, about which he coined the famous slogan "I liked the shaver so much, I bought the company"—"Procrastination is opportunity's assassin." If you suspect that a negative belief system is stopping you from achieving what you want in life, you need to find a way to change the system. Here are a few ways to go about making some positive alterations.

Convert Your Beliefs

Begin by creating a belief list. Take half an hour, and write down everything you believe about yourself, both positive and negative. Once you have the master list, separate the negatives from the positives. Perhaps you believe that you don't deserve that big house in the country or a loving partner. For each negative line item, ask yourself what keeps happening in your life that you'd like to change. For example, you keep getting dumped or you never have enough money.

At this point you have a choice. Either cross out the negative beliefs and replace them with positive alternatives, or better yet, convert the negatives into their positive counterparts. As an example of converting a negative, such as ending up loveless and alone, turn it into a positive, such as that a loving partner deserves you. If you adopt this tactic, make certain you have faith in your new belief. To reinforce the converted belief, list the reasons that it's true. The more effort you put into this process, the more likely you'll turn fantasy into reality.

Another way to turn a negative belief into a positive action is to alter the tone of the belief. For example, if you believe you'll never have enough money, think back to a time when you bought something in a store and the cashier gave you change. Obviously you had more than enough money on that occasion to buy what you did. So your belief of never having enough money isn't true after all. The belief then becomes that sometimes—and possibly frequently—you do have more than enough. If you believe that other people are untrustworthy, think back to the last time you were a passenger in a car or bus. You trusted the driver and the other drivers on the road, didn't you? Another distorted belief popped!

Limiting beliefs in particular have the power to alter your mindset so negatively that you give up on your own success before you even start. Suppose you're seeking a promotion at work. You've worked conscientiously, achieved your set targets, and been called a model employee. You feel that you've earned a promotion. Why *wouldn't* your boss consider granting it to you?

Supervisors are more likely to be in a good mood in the mornings, so when mid-morning arrives you straighten your tie and walk down the hall . . . just as a coworker asks if you've heard the news. The company has lost one of its largest clients. It's a *major* loss. Now you wonder if the timing for your meeting is wrong. What if your boss thinks you're partially responsible, that you didn't work hard enough on the account, that your department is at fault? He's certain to be upset, which won't help you get that promotion. Better to reschedule the meeting for later in the week.

Notice how easily you've shifted from a positive outlook to a negative. But you could turn the situation to your advantage and approach your boss with confidence, telling him that, had you been in charge of managing that account, the company wouldn't have lost the client. If he agrees to your promotion, you'll work harder than anyone to prevent a repeat loss. Besides, bringing you into the picture at the next level will show how seriously your boss takes the matter, which will reflect favorably on him.

So which is it for you, the negative narrative or the positive one?

Exercise

Your Brain's Bias

Does your brain prefer to seek success or avoid disappointment? The table below consists of a list of words that describe different feelings and emotions. Indicate the extent to which you feel a given emotion at this moment using the following scale:

1 = not at all   **4** = a lot

2 = a little   **5** = extremely

3 = moderately

Quality	Emotion	Score
+	Active	
-	Afraid	
+	Alert	
-	Ashamed	
+	Attentive	
+	Determined	
-	Distressed	
+	Enthusiastic	
+	Excited	
-	Guilty	
-	Hostile	
+	Inspired	
+	Interested	
-	Irritable	
-	Jittery	
-	Nervous	
+	Proud	
-	Scared	
+	Strong	
-	Upset	

Add the numbers of the positive emotions (active, alert, and so on, indicated by a + in the quality column). That number is your positivity score. Next, add the numbers of the negative emotions (afraid, ashamed, and so on, indicated by a − in the quality column). That number is your negativity rating. If your positive number is larger than your negative number, you're naturally a positive person—and vice versa.

To discover your positivity or negativity bias, subtract the smaller of your scores from the larger one. The larger the remainder, the larger your bias. To see your relative positivity or negativity as a percentage of your overall personality, add the two scores together and then divide each of them by the combined total.

Positivity score: _____

Negativity score: _____

Bias remainder: _____

Positivity percentage: _____

Negativity percentage: _____

Whenever a negative belief might be holding you back, you need to recognize it, deconstruct it, and rebuild it into a positive belief. Start by playing the Why, Why, Why? game. *I'm afraid my boss will yell at me.* Why? *My coworkers will believe that losing the account was really my fault.* Why? *Maybe they don't think I'm working as hard as I could be.* Why? Perhaps it's low self-esteem, a lack of self-confidence, or the thought of being criticized. Any one of those reasons could be paralyzing you into inaction.

To rebuild that negative belief, think back to a time when you showed a strong positive belief in yourself. (It's there—you just have to find it.) Once you identify it, visualize it fully until it becomes real again. At the peak of its intensity, acknowledge that, although you have demonstrated self-confidence in the past, something's blocking it now.

Next, list the emotional and material benefits of your promotion, followed by a list of the negatives. Do stronger positive outcomes dominate? Are there more positives than negatives? If so, you've just countered your brain's bias toward avoiding pain. Obviously, the benefits of the promotion far outweigh the risks of asking the boss for it.

By tackling these limiting beliefs one at a time, eventually you'll rid yourself of virtually every one of them and escape their shackles for good. The thing about beliefs and changing them is that you first must recognize when they're damaging. Then you must take action to change them. If you don't, they'll continue limiting you at every turn. If nothing changes, then nothing will change.

REMEMBER . . .

* Trust your emotions to help you achieve success.
* Monitor your decision-making process because your brain might be getting it wrong.
* Identify and positively realign negative beliefs.
* You must believe that you can change before you do.
* Whatever you learn can be unlearned.
* Don't give up on your own success before you even start.
* Whenever a negative belief holds you back, recognize it, deconstruct it, and rebuild it into a positive.

{39}

The Blame Game

HAVE you ever noticed how many people regularly lay the blame for their flaws and failings squarely at the feet of others? When people blame others, it's often because they're either too afraid or too ashamed to take responsibility for whatever has or hasn't happened. Psychologically, blaming others stems from a desire to protect our own emotions. But when we blame others, we forfeit our power to change.

Typically there are two different ways that people blame others, each driven by a different motive. We blame others when we are afraid to admit our own responsibility, and someone else carrying them can makes us feel more comfortable. This type of blame can represent a cry of pain or a plea for help. When a person desperately needs someone else to fix something, he or she may use blame to persuade others to take action.

A second way we blame others is more dangerous, more vague, and therefore often goes unnoticed. It's so powerful and effective that the person being blamed might actually accept responsibility for the grievance even if he or she shouldn't. For example, you tell a mutual friend that you think Michael is arrogant because he didn't accept your dinner invitation. Instead of first asking yourself why Michael didn't or couldn't dine with you, you disparage him for turning you down. Because you don't know why, you invent a reason, which results in you believing him to be arrogant. It happens every day. It's easier to blame someone else for your problems than to take action or address your own shortcomings. Lots of people fall into that trap without even realizing they've done so.

As a student, I didn't get good grades, and I attributed the fault to my teachers. When my bank balance was overdrawn, it was the bank's fault. (I'm still not sure how I came to that conclusion, but that didn't stop me from believing it!) The list went on and on. As you can see, I suffered from a victim mentality, which in extreme cases can become a martyr complex.

I blamed someone else—anyone else—whenever and wherever I could. Why? It was easier and more expedient than facing the problems myself.

How did I free myself from this victim mentality, and how can you? The first move I made was taking responsibility for my own actions. I realized that my life wasn't perfect and that my imperfections were my own doing. I had to stop blaming others and get a grip on myself before I could change. Here's how you can do it.

First, identify the source of your victim mentality. As with so many aspects of our personalities, a victim mentality tends to have its foundations in childhood. As a child, did you blame your teachers or siblings so your parents would be less harsh on you? Did they take your side? If so, they were feeding your victim mentality just as it was beginning to form. If you've ever been jilted at the altar, whose fault was it? Blaming the other party reduces your personal pain. The same goes for getting fired. Concentrating your anger and disappointment on somebody other than yourself eases your inner turmoil and protects you from the pain of the experience.

Once you locate the origins of your victim mentality, you can begin taking responsibility for your life. You can see how your next decisions in life are yours alone, bearing in mind that choosing not to do something or not make a decision still represents an action or choice, albeit negative. If a decision goes well, congratulate yourself emotionally. If not—and this is key—don't lapse back into the blame game and point the finger at yourself. Remember, there are no mistakes, only different outcomes.

You can transform the way you think both of yourself and your situation in life quickly and with minimal effort. Before long, it will feel natural to trash that childish victim mentality and take responsibility for your life. Once you do change, you may notice that you relate to other people differently. In my case, others used to influence where I went in the evenings, whether I would like a particular date, and even the type of car I drove. Wanting to fit into my various social groups drove me to seek the approval of others.

But then, as I took a greater stake in my own life, a totally unexpected shift happened. Instead of trying to fit in with people whom I didn't want influencing me anymore, those same people began trying to fit in with me. It was as if they too were looking for guidance and my coming-of-age was providing it for them.

Of course, we can't control *every* aspect of our lives. If you're sitting in your car, waiting for the traffic light to turn green, and someone rear-ends you, you're obviously not responsible for what happened, but you are responsible for how you react to the situation. Much of what happens or fails to happen to us is up to us as individuals. The bottom line is that we have to stop blaming one another and take control of our own brains, actions, and lives. Only then can we achieve the success we want.

REMEMBER . . .

* Blaming others is a symptom of a victim mentality.
* Choosing to neither make a decision nor act is still a decision or action.
* When you take responsibility for yourself, you become a stronger individual.
* You're not responsible for everything that happens in life, but you are responsible for how you react.

{40}

Success Breeds Success

I F those around you are negative—defeatist and resentful blamers and naysayers—then that's how your mind will make you feel. See for yourself. The next time you're in a group of coworkers and one or two express their negativity, watch the overall mindset and physiology of the whole group. All it takes is one or two negative comments about some imperfection, problem, or shortcoming to get the entire group thinking and acting negatively—which of course damages productivity.

This type of group activity doesn't happen just at the office, either. It also takes place at family gatherings, date-night restaurants, and elsewhere. The more you expose yourself to negative situations, the more likely you will become negative and the less will you strive toward success. So you need to learn to spot the telltale signs of those negativity hotspots. Typically these signs include an overall aura of disappointment and apathy, constant recollection of all that's wrong, and body language that communicates resignation and failure. When you find yourself among such people, you have one of three choices to make. The easiest is to blend in and become just as miserable as everyone else, but what good does that do? Who wants to be an integral member of a negative group destined for failure?

Deciding which type of social groups to associate with can have a profound impact on many aspects of your life. Think about the people closest to you. Are you working in a negative office environment or coexisting in a personal relationship that's going nowhere? Analyze your social situations often. Before long, you'll learn the signs of activities, groups, and even jobs with which you should or shouldn't associate. Success builds on success. Build a strong, healthy foundation by surrounding yourself with positive, like-minded people, and you'll find the motivation and drive to make it to the top and stay there.

So your second option is to aim to surround yourself with people who support your aspirations. Ask yourself whether those around you honestly share your desires and support your efforts and progress, or do they want to see you fail? It's easy to get stuck hanging around people who you thought were your friends just because it's convenient, rather than striking out on your own. Some people might want only to pull you apart, block your path to success, and prick holes in everything you do because jealousy or negativity reinforces their own shortcomings, inadequacies, and failures. It's always better to let go of people trying to bring you down than to drag out a relationship for old times' sake. The longer negative people dwell in your mind, the more they can derail your success.

For a positive example, take Roger Bannister, an English runner who broke the four-minute-mile record in Oxford on May 6, 1954. But then something significant happened. About seven weeks later in a race in Finland, Australian John Landy also achieved a sub-four-minute mile, beating Bannister's record by 1.5 seconds. Then, a year later, Hungarian László Tábori joined their ranks. They had all been training for years, but Landy and Tábori in particular had a powerful ally in that race. They believed that because Bannister had broken the record a year earlier, they could do the same—and they did.

So you can fall in line with the losers or align yourself with the winners. Your third and bravest option is to stay where you are but walk your own walk. Exude confidence and behave with conviction not as if you hope one day to become successful but as if you're successful already. It might feel a little strained at first, but before long you'll believe in the mindset and others will want to ride your coattails to success.

REMEMBER . . .

* Avoid negativity hotspots.
* Frequently take stock of your social situations and groups.
* Surround yourself with supportive people.
* Success builds on success.

Change the Script

I N the self-improvement genre, people often talk about positive affirmations. These are statements that declare something to be true. We use them to convince ourselves that we can reach if not exceed our goals.

The theory behind how affirmations work is that repetition codes them into your mind—regardless of whether they're true at the time you recite them. Repetition results in belief. Belief becomes perceived reality. Perceived reality creates self-fulfilling prophecy, and so the affirmation comes true.

That's the theory. In reality, the jury's still out on whether positive affirmation or self-motivation alone can bring about success. But a lot of scientific evidence indicates that we tend to limit our successes by limiting our beliefs. By using self-affirmation you might be removing mental stumbling blocks and limitations preventing you from reaching your potential.

We previously discussed how beliefs often take root in childhood and then program our psyches. They're powerful and can limit, damage, or counteract your intended success in life. Self-affirmation first drowns out the negative chatter emanating from distorted belief systems. Then, over time, the affirmation overwrites the negative beliefs, replacing them with positive thoughts that instill confidence, ambition, or whatever else you need to achieve your goals.

That all sounds pretty good, but just as affirmations constitute a vital facet of any personal-achievement formula, negative affirmations—denigrations, if you prefer—can do the opposite just as easily. If you tell yourself you'll never find your ideal partner or afford something you want, your good old gray matter will make that come to fruition. Your brain isn't all that judgmental, so long as you don't die, experience pain, or destroy all of humankind, it happily will help you fulfill what it thinks you desire.

Why? Your brain is more interested in filtering that endless litany of incoming stimuli into manageable chunks than it is in passing judgment on whether something is ultimately good or bad for you. If you fill your brain with positive dreams and beliefs, it'll work with them. If you stuff a bunch of negativity into your cranium, it'll do what it can to turn them into reality.

Here are a couple of examples of this principle in action. Salespeople use scripts called *closes*—rehearsed pieces of dialogue—to convince potential purchasers to buy a product or service. One is called the presumptive close. It works like this: The salesperson asks a number of questions to which the answer is always an obvious yes. "Does having a clean carpet appeal to you?" and "Would you like to spend less time vacuuming your home?" After a number of those comes the closing question, "If I could give you a clean carpet in half the time for only $5 a week, would that be a sound investment?"

The prospect, primed with a series of positives, naturally says, "Yes!" which is how positive affirmation can work in your brain.

On the flip side, many stores display window-front signs with messages such as "No shoes, no shirts, no service" or "No refunds!" Those messages prime shoppers to think negatively—and they do. They tend not to go into the store, or they go in but don't buy anything. Either way, you can bet your bottom dollar that they don't go back.

Here's another example that you've probably seen some version of in a shared post on social media or in an e-mail forward:

Aocdcrnig to rseecrah at Cmabrigde Uinervtisy, it dseno't mttaer in waht oderr the lterets in a wrod are. The olny irpoamtnt tihng is taht the frsit and lsat ltteer be in the rhgit pclae. The rset can be a taotl mses, and you can sitll raed it whoutit a pboerlm. Tihs is bucseae the huamn mnid deos not raed ervey ltteer by istlef but the wrod as a wlohe.

If you could read the sentences, your brain, confronted with near gibberish, successfully converted 280 disjointed letters into 69 words. Your brain makes sense of the nearly random letters by looking at them more as chunks.

So what does that mean? Stop focusing on what you don't want to happen and devote your attention to what you do want. We just saw how to make sense of a jumble of letters on a page by recognizing only a few of them. In the same way, when we listen to others, we tend to pay more attention to some words than to others. For example, we pay more attention to concrete words, particularly nouns, such as *car, road, cat,* and *tree* than less solid words, such as *very, not,* and *some.*

Again, so what does that mean? Let's assume a golfer stands on the course thinking, *Whatever you do, don't hit the ball into the water.* The first thing his or her brain does with that instruction is make sure the ball falls into the water. Why? Because it processed "ball" and "water" more effectively than "don't."

We hear more of the good in life than the bad, just as we'll hear more flattery than insults. As we go through life, we're constantly updating our self-image in that way. Call it selective hearing, but whatever the label, it keeps us ticking.

Cognitive neuroscientist Tali Sharot and his team at University College London conducted a series of experiments to understand more about the phenomenon. They discovered that the right side of the brain has this good-news bias and the left side doesn't. Their findings square with other research that has identified that the left and right sides of our brains are always arguing with each other. The right side of the brain conducts expressive and creative tasks such as recognizing faces and processing emotion, while the left takes on logic, language, and analytical thinking such as reasoning and numerical calculations.

All the science we've seen so far tells us, then, that right-brain thinking (emotional) is more powerful than left-brain rationality (reason). So it makes sense that, when we filter out so much information, we process the world around us more emotionally than rationally and with a positive bias.

This habit matters for people who always approach life negatively. When someone thinks, *I'll never find my ideal partner,* the brain filters and concentrates on the concrete words. What's left is *never partner.*

If you find yourself falling into that trap, you need to break the chain and start thinking in a more success-oriented way. One way to do so is to write down what you're thinking and how you're thinking it. Highlight each of the concrete words to see what message you're really sending to your brain. If the dialogue disagrees with your desire for achievement, don't change what you're saying but how you say it. In other words, spin it another way.

In the example of the golfer standing on the course, he or she shouldn't focus on the unwanted outcome (*Don't hit the ball into the water, don't hit the ball into the water . . .*) but on the wanted outcome. The golfer should concentrate on creating a clear, detailed mental image of the club striking the ball, which then lands in the middle of the fairway. When your brain undertakes a task, the more you heed the outcome you want, the more likely you are to get that success.

Here's a way to turn denigrations to your advantage by reframing what you're saying. Take the thought, *I'll never get a new job*, and rephrase it to, *When I get my next job, I will . . .* By moving the emphasis from negative to positive, you're priming your brain to seek a positive outcome. It's important to note here that the negativity limiting you is emanating *from* you. Once you understand that, you'll understand that you also have the power to change it.

Damaging negative thought patterns can limit our achievements by their sheer emotional intensity. As we know, only when events in our lives are tagged with emotion do they get filed in our long-term memories. Whenever specific emotions, especially intense ones, come into play, the experience becomes part of our memory. Along these lines, it's easier to recall negative memories when in a bad mood than it is in a good one. That's because we remember details and events in the state in which we first learned them. Whenever you're angry, you'll more easily recall other situations in which you were angry. So the less often you're in a bad mood, the less often you're likely to dwell on bad memories. Another good reason for thinking positively!

When you do find yourself thinking negatively, alter the tone of that negativity into something more positive or, alternatively, more focused. *Don't hit the water* needs to become *middle of the fairway*. Changing the focus changes what your brain will aim for.

Of course, part of whether you tend to think positively or negatively comes from sheer habit. If you've had a long history of negativity, you're likely to continue in that vein. Fortunately, you can break bad habits and replace them with good ones. Let's see how in the next chapter.

REMEMBER . . .

* Affirmations can counteract negative beliefs that limit our success.
* The brain cares more about filtering information than judging whether it's good or bad.
* When we filter the world around us, we tend to do it emotionally and with a positive bias.
* Stop focusing on what you don't want to happen, and concentrate on what you do want.
* We remember details and events in the emotional state in which we first absorbed them.
* We pay more attention to concrete nouns than other kinds of words.
* If you find yourself thinking negatively, don't change what you're saying but how you say it.

{42}

Making and Breaking Mental Habits

WHEN we think of habits, more often than not we focus on the undesirable. But habits can be both bad and good, physical as well as mental. What we have to learn is how to create good ones and how to remove those that have plagued us for years. If you give your brain the right set of instructions along with the correct motivation and emotional leverage to support them, it will do the rest.

The *American Journal of Psychology* defines habits as "a more or less fixed way of thinking, willing, or feeling, acquired through previous repetition of a mental experience." In physical terms, a bad habit is a negative behavior pattern acquired through previous repetition of a mental or physical experience such as nail biting, smoking, or overspending. But there are also good habits, which can be described as "a beneficial thought process or behavior pattern, triggered without conscious effort or even awareness." These include behaving courteously, saying please and thank you to others, and even brushing your teeth at bedtime.

It's the repetition of the thought process or experience that ingrains it in our everyday behavior, and typically a single stimulus will trigger the performance of it. For example, the thought of going to the dentist can cause a nail biter to bite her nails.

Many of our habits developed during our formative years, most often as a result of learning from our parents by observation and modeling. As we learn a new action, we create a neural pathway to carry it out. The more we conduct the behavior set, the stronger the pathway becomes. As we saw at the beginning of the book, our brains recognize a trigger that leads to the automatic execution of this behavior at the same time that the pathway is forming. So in the future the right sensory trigger initiates the habitual thought or action.

Knowing how habits form can help you take control. The first step toward removing a bad habit is awareness of it. Once you've focused on it, you can analyze the process. Think of a bad habit that you'd like to break. What's the trigger? Keep in mind that there may be more than one. Is it a feeling or something more physical? Is it a little of both? If your bad habit is biting your nails, maybe the trigger is the stress of driving to work in the morning. Now take the next step. Is it the act of driving itself or something deeper? Perhaps you're afraid of your boss. Once you identify the trigger, you need to rewire a pathway to overcome it. Unfortunately, that takes time, and how much time depends on how embedded the habit has become.

To rewire the pathway, you need to create a competitive response. Instead of biting your nails, get a stress ball that you can squeeze whenever you feel tense. Then, whenever you think of your boss, reach for the stress ball instead of biting your nails (though not while driving!).

Here's an example from my own life. I decided at a young age to become a professional jockey. As such, I had to keep my weight down. In those days, that meant smoking instead of eating. After my racing career ended, I decided to quit smoking. The trouble by that time, of course, was that I'd become addicted. In order to quit, I had to rewire. But how?

Since I had become a businessman, I spent much of my time flying. Nicotine patches, I discovered, reduced my cravings on long flights. I wrote down all the emotional reasons that I wanted to give up smoking and recorded myself saying them aloud. I played the recordings back as often as I could—in the car, on a plane, or even when walking the dog. The strategy was coming together. The nicotine patches worked, and my brain recognized emotionally why I wanted to quit. All that was left was the final push needed to quit on a particular day.

When I arrived home after a corporate retreat where I'd been doing team-building exercises, eating well, and enjoying time outdoors, the emptiness of my home struck me. My wife was away at the time, and so my own selfishness overwhelmed me, prompting me there and then to take more responsibility for my life and those I cared about.

On that particular Wednesday evening, I threw away my cigarettes, applied a nicotine patch, and quit smoking. I applied patches and tapped them every time I felt a craving, which was my substitute physical behavior in lieu of reaching for a pack. I also made a point of telling others that I had quit. Not wanting to admit failure to my peers gave me additional mental leverage.

The following Saturday, I went out to play a round of golf, and it rained a *lot*. A few times during the round I tapped my patch, addressing the momentary craving for a cigarette. When I returned home and went to take a shower, I noticed that the patch had fallen off somewhere on the course! But simply tapping where it had been had satisfied the craving.

In less than a week, my brain took what I had given it and transformed me from a pack-a-day smoker to somebody who hasn't smoked since.

Now what about good habits? How can you train your brain to embrace those? Whether it's eating a healthy diet, exercising regularly, or getting a good night's sleep, new habits first need to be learned and then wired as a neural pathway. Here's how.

Once you decide on a good habit to adopt, you need to create a step-by-step process to embed it in your mind. As I did above, first identify the emotional reasons you want to incorporate it into your life. Write down the reasons, say them aloud as you write them, record them, and play them back. Next, identify small, achievable steps that you can take toward the habit, and make them part of your routine. This also helps to create triggers that remind you to carry them out. In the beginning, triggers could be sticky notes in the car and on your desk or reminders on your smartphone calendar or in your email. Rehearse the behavior that you want to learn. At first, you'll have to make a conscious effort. But in time, depending on the emotional leverage you give yourself, it will become automatic.

Use the structure, logic, and reason of your prefrontal cortex to give your fallible, overworked brain the emotional help it needs to achieve your desired success.

REMEMBER . . .

* Habits can be bad or good, physical as well as mental.
* Sensory triggers cue habitual thought or action.
* Create competitive responses to rewire the pathways of habits you want to break.
* To create new, good habits, identify your emotional motives, establish small steps to incorporate into your routine, create and use triggers, and rehearse the behavior set.

{43}

A Goal Needs a Roadmap

WHAT does success mean to you? What do you really want? What are your goals?

Those may seem like simple, obvious questions, but you need to think about them carefully. We've talked a lot about how your brain works, how to harness the power of your emotions and prevent them from holding you back, and strategies for overcoming social obstacles. But the single most powerful factor that prevents people from achieving success is their own lack of clarity about what they want.

Most people want money. But from the brain's perspective, money is a flawed goal. As we saw earlier, money itself isn't the target, it's the means toward what you really want. It's all well and good to want a million dollars in the bank, a five-thousand-square-foot house in the suburbs, and a swimming pool the size of New Jersey. But why? Your brain evolved long before any of those possessions ever existed. It fundamentally can't grasp the concepts of high-yield accounts or country estates.

The real goal might be to give up work and spend more time at home or more time exercising in the pool so that you feel better about yourself physically and emotionally. Either way, the money is merely a means to an end. Your brain can't achieve a method—even the sentence doesn't make sense! Let's quickly play the Why, Why, Why? game to understand how flawed thinking prompts us to misinterpret our goals and then misdirects our brains to go after the wrong dream.

I'd like a million dollars in the bank. Why? *Because I don't want to spend sixty hours a week in the office working to support my family, whom I then never see.* Why? *Because I'd like to spend more time with them and I'd like to be under less pressure.* Why? *Because if I lose my job, my family will think I failed them and they'll resent me for never being around.* Here, the underlying goal is wanting to remove the fear of failure and resentment rather than having a seven-figure bank account. But removing the fear of failure is an emotional goal that your brain can help you accomplish.

A goal is the end result of expended effort. The key phrase there is "end result." However you define success, you need to create your roadmap from the bottom up, not from the top down. When setting goals, you need to think through them carefully to help your brain fully understand what you're seeking. Otherwise, how can you know when you've reached your goal, and how can you create checks and balances to make sure you're on the right track?

So it's no good starting with a list of material possessions. Begin your list by letting your brain know what your success feels like emotionally. Once you've done that, you can associate those emotions with the material world. You can use objects and possessions as mile markers on the road to reaching your emotional goals—as long as they don't serve as the goals themselves.

Exercise

Profile Your Goals

Remember the three components of emotion? Your brain uses enjoyment, excitement, and intimidation as the mental switches to prepare your body for instinctive action. Here's a simple tool that will help you understand your emotional connection to a particular objective. Write a specific goal—a person you want to get to know, an object you desire, or an event important to you—here:

Now, let's look at how you feel about the components of the objective, where 1 is not at all and 10 is extremely.

1. How happy do you feel about trying to attain that goal?

   1   2   3   4   5   6   7   8   9   10

2. How excited do you feel about trying to attain that goal?

   1   2   3   4   5   6   7   8   9   10

3. How scared do you feel about trying to attain that goal?

| 1 | 2 | 3 | 4 | 5 | 6 | 7 | 8 | 9 | 10 |

Now imagine how you will feel when you have achieved that goal.

4. How happy will you feel once you've reached the goal?

| 1 | 2 | 3 | 4 | 5 | 6 | 7 | 8 | 9 | 10 |

5. How excited will you feel once you've reached the goal?

| 1 | 2 | 3 | 4 | 5 | 6 | 7 | 8 | 9 | 10 |

6. How scared will you feel once you've reached the goal?

| 1 | 2 | 3 | 4 | 5 | 6 | 7 | 8 | 9 | 10 |

You've just created an emotional profile of what reaching that goal means to you. Now let's compare your responses. Transpose your answers onto this table, and then we'll look at what the results mean.

The goal:	How you feel about trying (1–3).	How you'll feel when achieved (4–6)
Enjoyment		
Excitement		
Intimidation		

First, which is the larger of the enjoyment numbers? Usually it's how you'll feel once you have attained the goal. If not, then you're not emotionally serious about achieving it, and you need to work on your emotional motives.

What's the largest number you gave to how you feel about trying to attain this goal? If it's enjoyment, you're looking forward to the challenge. If this is the case, you just need the

leverage to push you into action. Stay focused on the pleasure or happiness that you'll reap.

If it's excitement, you're emotionally heated, which is a
good state for determination—but be mindful of the impact
that your anger and aggression can have on others. Success
can be lonely if you can't share it with anyone.

If the largest number is intimidation, then you're scared
to tackle the goal. You need to address this fear head-on,
or your brain will dodge the necessary steps. Take the time
to understand why you're afraid, and then reframe that fear.

Let's take Charlie for example. Charlie has tried numerous strategies and
techniques to make more of his life: fad diets, motivational speakers, and
self-help books. Still, the success that Charlie has achieved has been hit-or-
miss. Now he's ready to get serious.

Charlie played the Why, Why, Why? game and discovered a couple of
significant emotional drivers. First, he wanted to be loved, and second he
wanted to be respected. Both of these emotional states differ from land,
money, and property. Love and respect are what Charlie ultimately wanted.

So he took a blank sheet of paper and set about creating a master plan.
He divided the page into two vertical columns and wrote *love* at the bottom
of one and *respect* at the bottom of the other. Those were his goals. Then he
closed his eyes and visualized how love and respect felt to him.

For love, he imagined relaxing on a sofa with the woman of his dreams,
their two children safely asleep upstairs. Focusing on the sofa and the
two-story house, he wrote both of those words on the list above *love* along
with emotional descriptions of both. He also envisioned teaching his children to swim. Focusing on that activity, he wrote *swimming pool* in the
love column.

When it came to respect, Charlie imagined people whom he admired
coming to him for help and advice. He wrote *source of knowledge* in the

respect column. He envisaged listening to a conversation where he overheard his work colleagues commenting on how much they admired him. For this, he noted *work colleagues*. He also pictured his parents proudly introducing him to their peers and describing him as an expert. He put *parents* into the respect column as well.

Over a period of several days, Charlie generated a list of nearly twenty material objects, each of which had strong emotional ties to potential achievement. Then he prioritized the lists by which emotional mindsets (via material possessions) were quickest and easiest to attain. For example, buying a good sofa was simple, as was buying and reading a couple of books on his profession to make him more of an expert. At the end of the process, Charlie had identified the emotional states that he wanted to achieve, attainable milestones to mark his progress, and the payoff for attaining each goal. He had given clear and comprehensible information to his brain about what he wanted to achieve in terms of how he wanted to feel. Now he had a plan.

That same exercise will help you, as it helped Charlie, to identify where you want to go. But setting goals means facing reality. How long will it take you to reach a particular goal? If you're in your late twenties and one of your emotional goals is the pride you'll feel when dropping your children off at college, then obviously you've got a ways to go. Don't discard that long-term goal, though. Break it down into its component pieces: the joy of finding a spouse who loves you, the euphoria at the birth of your children, the gratification of helping them do well in school, the triumph of learning what colleges accepted their applications, and so on. Now your brain has simpler and more manageable short-term tasks.

Earlier in my life, as I was setting up my business, I had several goals, one of which was to make a lot of money. But I soon developed a strategy to think more concretely. How long can I support my family and myself without another trade order coming in? At first, I aimed for only a few days. In time, that span grew to several months. After trading for just one year, I had enough savings to support my family for twelve months.

My brain was pushing me to create wealth at more than twice the speed I needed it!

Track Your Milestones

How can you tell if you're succeeding or failing at your goals? Here's a tool that you can use to measure your success along your journey.

We've talked about how our brains often distort past events. They remember the good and try to forget the bad. But even then, they tend to distort both. The same can happen on your road to success. You should regularly evaluate where you stand and accept that you'll need to change course occasionally. Remember, you might not always get the outcome you are expecting.

Your brain has phenomenal powers of providing you with inventive and effective ways to reach what you crave, but it's not exactly objective when it comes to evaluating its own performance. To overcome that shortcoming, you'll need to keep a success journal or some kind of record of your progress. Keeping a journal will help improve your effectiveness and the likelihood of reaching your goals.

Writing events down stores them more effectively in your memory. It also provides you with the opportunity to analyze them the following week or the following year. Also, itemizing what you've achieved and what you didn't accomplish will help you discover what works for you and what doesn't. Your success journal will also remind you of what you should have done but didn't.

We all lead busy lives, so you don't have to write in it every day. Be realistic about the time you can and want to devote to tracking your progress. Set aside half an hour one day a week or one day a month to track your progress, but stick to that schedule religiously. There's no point in having a success journal if you never use it!

But what should you write in it? First, generally note how the journey's going. How satisfied are you with your progress? What reasons or excuses do you have for why you might be off-track? Note any successes you've experienced, no matter how small. As we've seen, that will help train your brain to focus on all of your successes. Also write down what's gone wrong or what you still haven't done. Explain what caused the shortcoming and how you'll prevent it from happening again. By focusing on preventing recurrences, you're replacing the negative memory of the unwanted outcome with the more positive thought of future improvement.

To get the best effect, give your goals specific dates in the future. That will help your brain establish a timeline and sense of urgency for each of the items you want to accomplish. But again, be realistic with your estimates. Some of your goals may take weeks to achieve, others months or even years.

Keeping a journal is a good way of tracking how well you're doing, but you need to read it and review your progress regularly. That will help you refine your strategy while also preventing you from doing the same thing over and over but hoping for a different result each time. Write down your goals, memorize them, and share them with others. Tracking them in your journal will give you an excellent source to refer to when you need a reminder of what you're aiming to achieve and when family, friends, and colleagues ask how you're doing.

REMEMBER . . .

* Be clear about what you really want.
* Use material objects as milestones on the way to success, not as the objectives.
* Create your goals from the bottom up.
* Break long-term goals into smaller components.
* Keep a detailed success journal to track your progress.
* Set specific dates for reaching your targets.

{44}

Visualize Success and Failure

B Y now, you should know what success means to you in terms of how you want to feel in the future. But let's go one step further. If you want real success, you need to *need* it. It's not good enough merely to want it. Your brain has to recognize that success is an imperative.

For many people, love is a powerful success goal. Assuming it's one of yours, you need to create different mental scenarios in which you experience the emotion of love. For example, in the last chapter Charlie envisioned cuddling on the sofa, which acted as a trigger to feeling love. You need to create some triggers of your own. What about walking hand-in-hand with someone you love? How about sharing a romantic dinner in your favorite restaurant? If you've already created some, add some more. The more numerous and more powerful the mental representations you create, the greater influence they'll have on your brain.

Next, you need to try something that sounds easy in theory but can prove difficult in practice. You need to set aside time to visualize your success. Aim for the same time each day, and devote at least ten minutes to yourself in absolute privacy.

Years ago, when I was first putting the principles of this book into action, getting away by myself was challenging. Not to be deterred, I resolved to leave the office each lunch hour and drive to a secluded spot a mile away. There, I enjoyed half an hour of quality personal time, concentrating on my goals. After only a few days, just sitting quietly in my car gave me the emotional sensation of the success that I wanted. As a result, my brain cued that feeling not only during my lunch hour but also whenever I got into my car. Driving to business meetings acted as a trigger to thinking of my impending success.

But the important part is that, however hard it is to find time to think, you can always find the time. Lots of people use lack of time as a catchall excuse for not doing what they want or what they should, either at work or home. Amusingly, those are often the same people who can wait in line for ten minutes at their local coffee shop for that morning latte!

––––––

Envisioning success moves us toward achieving it. But what about the opposite? What about envisioning failure? Here's a game I'd like you to play. Take one of the goals that you've set for yourself, and imagine you're failing at it. Feel the negative consequences of not achieving what you want. What's the emotional pain you're experiencing? Pay particular attention to that, and make sure you differentiate it from the emotional feeling of success.

How do you imagine your failure mindset? Do you see yourself sitting at home, alone, on a Saturday night? If so, how does that feel? What about seeing others cooking Thanksgiving dinner without you? How would it feel to see someone you admire take center stage at a party while you silently hug the wall?

Whatever you do, though, don't focus on the negative as often as spending quality time visualizing the positive outcome of achievement. It's a mental exercise that you should use every once in a while to help your brain stay focused on the good while avoiding the bad. If you have a positive focus session each day, once a month is sufficient for the negative emotion exercise. The purpose of feeling the negativity is to reinforce that failure is more painful than success. Remember, we always seek to avoid pain more than to attain pleasure. By reminding our brains of both options, we're reinforcing which way to go.

Now, let's add more components to creating your mindset for success.

––––––

REMEMBER . . .

* Help your brain recognize that success is an imperative.
* Set aside quality time to visualize success on a regular basis.
* Don't be afraid to imagine failing.
* Remind your brain that failure is more painful than success.

{45}

Make It Real

ARLIER, we discussed associated and disassociated states of mind. In an associated mental representation, you're inside the image or event, looking out through your own eyes as if acting in a play or a film. In a disassociated mental state, you're outside the event or picture, watching yourself and your behavior as if seeing yourself on a screen. But before we see how to use these states of mind, let's do a quick recap.

Of our two forms of memory—short-term and long-term—we know short-term memory becomes long-term in one of two ways. The first is through repetition, and the second is through tagging the memories with emotions. You remember some school friends more than others because you liked them or you hated them, for example. When a long-term memory reactivates, the emotional processes that created it also reactivates. Since this is partly a chemical reaction, the same chemicals created during the experience are created again when your brain recalls the memory from long-term storage.

Each time you recall and relive a memory, it activates and strengthens the neural pathways in the brain. If the memory of a situation has become distorted over time, the chemicals released by the brain can change in the same way the neural pathways change. As a consequence—and this is the payoff—the memory can become either strengthened and more meaningful or weakened and less intense. That means that we can change or even eliminate their emotional impact. In other words, we can't change the past, but we can change how we feel about it.

We saw in the screen-compactor exercise (chapter 19) how to dissociate a negative memory to weaken it and strip its credibility. That was a way to undo damage, but we can use the same technique in reverse to reinforce positives.

Take a specific goal that you want to achieve. Begin in a disassociated state by standing on the metaphorical sidelines, and create a visual representation of the goal. Charlie, for example, would see himself on the couch with his wife or in the pool with his kids. As you behold the scene, keep adding more visual details, including color, background imagery, and so on. Make it as realistic as possible. Now animate the image. Create a moving picture, and pay attention to the motion of light, people, and objects in the image.

Once you've added as much visual detail as you can, add sound. Again, include details of what's being said in the foreground along with any background noises. Next, bring in any smells that further embellish your representation before finally incorporating any associated tastes.

Once you've done all that, stand back and experience the whole scene playing out. Involve all of the senses that you've harnessed. Go through the event several times before taking the final, most powerful step.

As you rehearse your mental representation again, step into it this time and become fully immersed. Change your point of view from outside looking in, to inside looking out. Add to the experience what you can feel physically, what you're sitting or standing on, what you're touching with your hands. Now you're literally in your own head. Play the starring role, and experience the event through all your senses.

Be sure to invest the time and effort necessary for developing this ability to switch between associated and disassociated states of mind. By doing so, you're communicating with your brain both emotionally and chemically, and you're the only person in charge. This is the power of psychology.

REMEMBER . . .

* We all have the power to strengthen or weaken our emotional memories.
* Develop the ability to switch between associated and disassociated states of mind.

{46}

Reframe Failure

AS we saw, one of the surest ways to prevent success from happening is by believing it won't happen. If your brain contains any negative or limiting beliefs that are more intense and powerful than the positive alternatives, you're not going to get there.

As we go through childhood and into adulthood, we learn what to believe. Unfortunately, certain events or factors can misguide us. Being abused as a child or by a partner can lead a person to believe that the world is a bad place and everyone in it is terrible and vindictive. As a result, the person might believe that he or she is unlovable and can't commit to loving anyone ever again.

The good thing for all of us is that beliefs are only that—they're what we believe. If we cease to believe them, they cease to hold true. We can change them.

The global market–research company Ipsos reported that around 10 percent of the global population believed the world would end on December 21, 2012. But as we all know, it didn't. The forecast came from the Mayan calendar, which never actually said the world would end but rather that their Long Count calendar, carved in stone some six thousand years ago, ended and then started over.

Lots of people all over the world got frantic about this ancient Mesoamerican "end-of-time" event. Russians stockpiled food and water. Italians dug underground bunkers. Cynics with a practical streak placed bets with bookies. But when the winter solstice of 2012 came and passed without incident, we were all still here. It was a less technological repetition of what happened on January 1, 2000, as unfounded Y2K fears dispersed.

But how did those passionate calendar-watchers explain their misplaced beliefs in 2012? Some blamed an unsubstantiated calculation error and claimed the *real* event still lay five hundred years in the future. Others insisted that disasters, such as the Tōhoku earthquake and tsunami in Japan on March 11, 2011, proved the prediction true, making excuses for the final date not aligning as expected. Even the most ardent believers changed their beliefs when December 2012 came and passed without incident. Moral of the story? However deeply held a belief, it can still be changed.

Let's revisit the list of beliefs that you created in chapter 38, "How Belief Systems Form," and audit whether any of the negative beliefs are still influencing your goals. First, you have to recognize the negative belief for what it is, which not everyone can do easily. Do you expect to find true love, or are you planning to live your life alone? Do you see something you'd like to own and automatically think, *One day soon*, or do you sigh and think, *Never going to happen?* Many people prefer to regard even damaging negative beliefs—such as excessive self-sacrifice or working slavishly without end—as of some cryptic benefit.

The second step is to reframe the belief as a positive. For example, the original belief may have been *I'm too old to learn a new language.* You can recreate the belief in a more positive context, *I haven't been able to learn Spanish yet, but I'm going to do it.*

Forming positives from negatives takes a lot mental effort. If you get tired, take a break to recharge your brain for another session. Eventually, you'll do it. As you repeat the process, you'll notice yourself developing a more positive outlook of the world.

Reframing your whole belief system also takes a lot of time and effort. Think for a moment about how many years it has taken to develop your current belief system to the level of power that it has. The key is to seek out small ways in which your beliefs are changing and then focus on them. If at first you can't play the piano, learning where middle C is and hitting that key means that now you can play a little bit of piano. You're a long way from mastering Chopin's *Etudes*, but you're not a total failure. Now add a second note and a third, try some basic chords, and then use both hands. Each time you advance, focus on what you've learned, not what you haven't. Playing a single piano key can change you from being a failure to achieving success—as long as you think about it that way.

REMEMBER . . .

* If you believe success won't happen, it won't.
* All beliefs can change.
* Reframe negative beliefs into positives.
* Focus on your overall outlook to help handle stumbling over individual instances of failure.

{47}

Adjust Your Approach

NARCOTICS Anonymous famously describes insanity as "repeating the same mistakes and expecting different results." Sometimes our brains don't learn, and after a mounting number of failures we simply give up. If you bake a cake for fifty minutes and it burns to a crisp, you know to bake it for less time or at a lower temperature on your next attempt. You don't keep trying the same baking time and temperature over and over again. But too many of us do just that, replicating the same old methods that failed to get results in the first place.

Success is attainable once you discover the appropriate steps to achieving it. Don't give up the chase, but do prepare to give up the approach. Each time you get an unexpected outcome, apply what you learned from that attempt, adjust your approach accordingly, and try again.

Here's an example for those who have children. How long did it take you to teach them to walk? After how many attempts did you give up and admit that they would never walk? Unless the child has a disability, no one ever gives up on teaching a child to walk. You may have had to fine-tune your approach, but eventually your little one learned.

Achieving your personal goals is no different. If you're trying to lose weight or quit smoking and you succumb to a bar of chocolate or a cigarette, don't call it a total failure and give up—it's just a temporary setback. You abstained successfully for several days, so your brain knows how to do the job. It just needs to learn to do it a bit longer.

Before changing your unsuccessful approach, isolate why you didn't succeed the last time. Feed that information into your achievement strategy, and try again. It won't take long to consign those chocolate bars and cigarettes to the past.

Also, the sooner you realize you've slipped, the less severe the slip-up needs to be. After you light that cigarette and pull the first drag, *stop*. If you eat a couple of cookies, don't beat yourself up, but don't eat the whole bag, either! Don't wait until you've smoked or eaten the whole thing, or you'll be rewarding your brain and training it to think that disobeying your emotional desires is chemically rewarding. You should be associating the lapse with negative emotions and mental pain.

This strategy also works when job hunting. You see lots of listings and keep sending out those resumes, but to no avail. You may land a few interviews, but then no callbacks. You don't want to keep wiring into your brain an ineffective neural pathway that says the approach is just right, that you're doing the best you can. Instead, set your brain to working harder. Every résumé you send out should be different and better than the one before. Each interview you attend should be more effective than the previous one. That's the message to send to your brain. Keep striving to better yourself each time.

People frequently complain about their résumés not being strong enough or their interview techniques lacking polish. If that's you, it's time for a change. Stop grumbling, and ask a peer or, better yet, a mentor what's wrong with your approach. Pull key language from the jobs you want to have to describe what you're doing now. Practice interviewing with family or friends. Keep honing your material until it's top-notch. Keep practicing until you succeed. Simple, isn't it?

Well, no.

Remember, the human brain has evolved over millions of years. By now, yours is pretty well set in a number of ways. It tries to protect you from pain at every opportunity, which includes the mental pain of rejection. Without knowing it, the strategy your mind employs to help you get an interview might actually be protecting you from being rejected. If your résumé doesn't land you an interview, then you can't be rejected, and if you can't be rejected, your brain is happy. But are you?

To avoid that mental trap, take a tip from James Dyson, designer of the popular Dyson line of vacuum cleaners. "I made 5,127 prototypes of my vacuum before I got it right. There were 5,126 failures. But I learned from each one. That's how I came up with a solution. So I don't mind failure. I've always thought that schoolchildren should be marked by the number of failures they've had."

As Nobel Prize–winning author Samuel Beckett once put it, "Try again. Fail again. Fail better." Never stop, never give up. Keep refining and revisiting until you've reached your goal. It takes hard work, but you'll get there.

———

Sometimes it's not the approach that's wrong. In some cases, incomplete knowledge leads to error. We often make decisions on subjects about which we don't have all the relevant facts. Poor communication also plays a role in human error. Sometimes we don't ask the right questions to allow us to come to the best decision. Changing conditions may play a role. A correct decision today can be dead wrong tomorrow. Governments are particularly vulnerable to that trap because of the time it takes to write, vote on, and enact various laws and other acts (not to mention that most politicians are more concerned about short-term re-election than long-term decisions). Pressure to perform also causes mistakes in judgment. We tend to commit errors when stressed or when we have to make decisions too quickly.

After a list like that, you might think it's impossible to do anything right in life. Not true, but we can help our brains make fewer mistakes. All too often most of us let emotional and mental habits control our decision-making processes, which almost always leads to thwarted progress or success. Instead of making decisions purely to address the frustration, consternation, confusion, or confrontation we're experiencing, think in terms of what decision makes the most difference to your values and to what's important in your life. When you think in this way, you'll make better decisions and fewer mistakes.

Take, for instance, somebody bumping into you at the supermarket. You instinctively feel anger because you're reacting out of habit. The same is true when a pedestrian or driver cuts you off. Turn that anger into something more constructive by reframing your reaction and embracing a more positive response. "What a sad life that selfish, careless person must lead. I'm so glad that I'm more self-aware and considerate." It might sound simplistic or even hokey, but doing that has helped many people in all walks of life to attain greater personal achievement.

Here's another example. Somebody I know tended to let life overwhelm him. All was good for a while, and then suddenly, for no apparent reason, he couldn't cope. He let his inbox overflow with hundreds of e-mails while becoming powerless to perform his job. A snowball effect resulted. The more he fell behind, the less he could accomplish and the more over-whelmed he felt.

After a couple of talk sessions, he learned to understand how his brain worked and completely changed his outlook. To help that change, he wore a rubber band on his wrist that, whenever he caught himself slipping back into a habit-related choice, he snapped against his wrist—*hard*. It hurt, but it worked. In only a couple of weeks he transformed his decision-making process, stopped falling victim to the snowball effect, and today reaps greater success in life. Whenever his brain slips into habit-driven mode, he snaps the rubber band. His brain responds by considering alternative responses.

Having looked at why we're prone to mistakes and what we can do to make fewer of them, let's dig a little deeper into mistakes that really aren't mistakes at all.

———

One of your goals in life is to work toward achieving the best results possible, not playing all-or-nothing roulette with your decision-making process. Here's one way to take the result from a bad decision and convert it into something better.

Think of a life mistake the way you would about practicing a sport such as baseball or basketball. If your first free throw falls to the left of the basket, aim your next shot a bit to the right. If the ball drops short of the catcher, aim higher and throw harder. All of us come up short from time to time, but if we miss a throw we don't give up the game forever. Instead, we refine our techniques in search of a better outcome.

A couple of friends identified a gap in the business-presentations market a few years ago. Through research they found that businesses create millions of PowerPoint presentations every day. Furthermore, their research showed that many of the presentations weren't very good, which comes as zero surprise to anyone who has ever suffered through one.

My entrepreneurial friends used that information to find a business to train people on how to make presentations more effectively so that audiences absorbed more of what they were saying. They wrote a book, created more than seventy training modules, and developed an entire marketing campaign, which included an informative, interactive website and a comprehensive PR strategy.

The book was published at the same time that the PR team executed the campaign to raise awareness about the subject and the book. But in the first month, they sold only six books and not one training course. It was a disaster. They easily could have given up, closed up shop, and slithered back to their old jobs. But they didn't. Instead they realized that they had succeeded in a small way. They had sold six books to the public. The next step to increase their success was to work on the numbers.

They tweaked and developed their concept. They diversified into the education market, showing teachers and lecturers how to embed their knowledge more effectively into the minds of students. Then they expanded operations into other countries. What began as a disaster soon blossomed into a new way of looking at the market for their product. You can similarly reduce the number of mistakes your brain allows you to make by thinking of them as refinements.

By viewing the unexpected not as a failure but as an unanticipated result, you'll find yourself less hesitant about trying different strategies. Your fear of failure will diminish as you realize you're experimenting toward achieving success and not recklessly gambling with the future. Concentrate less on the mistakes you make and more on creating alternative outcomes, and your attitude and achievements will soar together.

Will you never make another mistake again? Hardly. But adjusting your approach opens new doors, new avenues, and new ways to succeed that you never thought possible.

REMEMBER . . .

* Never give up the chase, but prepare to give up the approach.
* Apply what you've learned from unexpected outcomes and try again.
* Don't let small slips become huge setbacks.
* Keep refining until you've reached your goal.

Hypnotism?

WHEN you see or hear the word *hypnotist*, you probably envision a campy stage performer wearing a long black cloak and swinging a shiny pocket watch. That popular representation, however, bears little resemblance to actual hypnotherapy. People undergoing hypnosis aren't half asleep but very much awake. They retain total cognizance and absolute free will. Whether anyone can use hypnosis to achieve success remains open to debate. Either way, you must understand how it works, or doesn't, before hoping to derive any benefit from it.

Hypnosis is a trance-like state. In that state, a person is extremely relaxed, enjoys a vivid imagination, and is susceptible to suggestion. Typically, someone hypnotized blocks many of the stimuli around him or her while focusing intently on a specific subject. Under hypnosis, people blissfully tune out their day-to-day cares and worries while feeling relaxed and uninhibited. Think of it like the state you enter while reading a good book. The text hypnotizes you to shut out the real world around you. You involve yourself mentally with the characters and plot in the story so much that you distance yourself from your cares and worries. In that state of self-induced relaxation, you find yourself more imaginative and open to suggestion.

Now, does hypnotism work? Yes, it really does. According to Stanford University's David Spiegel, one of America's leading psychiatrists, it has an impact on the brain that can be measured scientifically. For a report to the American Association for the Advancement of Science, he scanned the brains of volunteers who were told that they were looking at colored objects when, in fact, the images were black-and-white. A scan showing areas of the brain that register color displayed increased blood flow, indicating that the volunteers "saw" the images in color. As Spiegel told the delegates, "This is scientific evidence that something happens in the brain when people are hypnotized that doesn't happen ordinarily."

But why? Here's one hypothesis.

The human brain consists of a small, relatively limited conscious aspect, along with a much more powerful subconscious. The conscious part helps us make sense of the world around us. The subconscious handles everything else. It controls the autonomic processes that you don't think about, such as increasing blood flow to the muscles, decreasing blood flow to the surface of the skin, and breathing. It's where our thoughts, memories, and accumulated experiences reside. It also controls our emotions, habits, and responses to the world around us.

When the conscious mind can't manage its task load, the subconscious jumps in to help. When the subconscious is making sense of a conscious problem, both parts of the brain are working together to find a solution. During such times, channels of communication not vetted by the more reasonable conscious mind open. In effect, hypnosis works by bypassing the conscious mind and speaking directly to the subconscious. It updates the subconscious mind with new information, similar to reprogramming a computer while bypassing a firewall.

The left brain is the logical, methodical, language-oriented half of our minds, while the right brain is more creative and intuitive. Occasionally, the noise of the logical left brain drowns out right-brain creativity. Some scientists believe that hypnotism gives the right brain a louder microphone while the concentration and relaxation of hypnosis temporarily quell the constant chatter coming from the left brain.

Of course, not everyone can be hypnotized. If someone fails to enter a trance, it's usually because he doesn't trust the hypnotist, won't fully relax, doesn't go with the flow, or is interrupted during the induction of the trance.

Hypnosis and hypnotherapy are closely related, but they do differ. Hypnosis is the relaxed state of consciousness itself. Hypnosis as used by professionals is not to be confused with stage-variety hypnosis, which is entertainment rather than therapy. Hypnotists induce hypnosis, although the occupational term sometimes has bad connotations because of the activities of stage hypnotists and charlatans.

Hypnotherapy, on the other hand, uses hypnosis as part of a therapy or treatment to help individuals improve their lives, whether for stress reduction or other medically or emotionally related issues. Hypnotherapy isn't a substitute for professional medical care, but it does act as a complement to it. A hypnotherapist who provides hypnotherapy usually has additional professional training, certifications, and qualifications in therapy as well as counseling skills. The hypnotherapist usually provides in-depth consultation prior to the hypnotism part of the therapy.

The most important aspect of successful hypnosis is complete relaxation. Much of the induction process aims at clearing the subject's mind, relaxing, and distracting the more logical and rational parts of the brain. The hypnotist asks the subject to visualize walking down steps or to count down as part of the induction. Once the subject crosses into a trance-like state, it's possible to convey messages directly to parts of the brain normally off limits to verbal suggestions. In this state, the brain often accepts messages such as "You are in charge of your life and no longer need to rely on toxic cigarettes." A complete behavioral turnaround, however, requires a number of sessions to remove all associations, depending on the mentality of the subject.

A hypnotist placing you into a hypnotic state has you close your eyes and relax. You need to feel completely comfortable as you listen to his or her instructions. As you become increasingly relaxed, the hypnotist has you focus on different parts of your body, beginning with the top of your head, moving steadily downward, one body part at a time. This element of the process helps you relax while at the same time taking your mind off the hypnotism itself.

As your level of relaxation deepens, the hypnotist has you focus intently on the visual aspects of a calming mental representation, like a walk along the beach or a pleasant snowy night. After that, he has you move your attention to what you can touch and feel, utilizing all your senses until you become totally absorbed in your own mental representation. At that point, you become receptive to key messages.

As your relaxation nears completion, your breathing slows and your facial expression relaxes. The hypnotist addresses the reason for the hypnotism. Ideally, he or she directs you to reach your own conclusions about why you should change a particular aspect of your life. Once he or she has prompted you to embed the messages in your subconscious (such as "Too much alcohol is self-destructive"), the hypnotist gently brings you out of the trance by having you count slowly backward from five to one. With each number, you begin to "awaken." When you reach the number one, the hypnotist tells you that you are wide-awake, full of energy, and feeling great.

Hypnotism, then, is a powerful auxiliary tool that can help you lose weight, quit smoking, improve sports performance, or strengthen your self-esteem and confidence. It can also help reduce stress and manage other strong emotions, such as anger, guilt, or sadness. But remember that hypnosis is powerful stuff. Don't try it without professional guidance.

However, there is a variation of the hypnotic technique that you can try on your own to help you train yourself to reach a particular goal. Here's how you can harness the tenets of hypnotism yourself.

Following the hypnosis process above, write a relaxation script that includes what you want to change in your life and record yourself speaking it. Play the recording and listen to your own voice telling you to relax and focus on what you want to change. Your mind hears you instructing your own subconscious. It's a fast-track tool to help you make those dramatic changes in your life.

REMEMBER . . .

* Hypnotism does affect the brain.
* Hypnosis is a state of relaxed consciousness.
* Hypnotherapy uses hypnosis to help people improve their lives.
* You can harness the power of hypnotism to tell your own brain what you want.

49

Anchors Away

A NCHORING describes the act of associating a specific behavioral trigger, such as snapping your fingers or tapping a particular part of the body, with a particular mindset. Anchors provide links to emotional states, and you can create them as subtle switches or conscious behaviors such as the routines athletes use to get "in the zone" before competing. Once you've mastered the art of anchoring, you can make powerful changes in your life as well as influence other people.

In Russian physiologist Ivan Pavlov's famous experiment with dogs, he sounded a bell (conditioned stimulus) as he gave the animals their food (unconditioned stimulus), which caused them to salivate (unconditioned response). After a number of feedings, he caused the dogs to salivate (conditioned response) simply by ringing the bell—without introducing any food. This became known as classical conditioning, and it's how anchoring works.

Hypnotists and psychologists use anchors, but they also occur naturally in our everyday lives. For example, we are anchored to stop at red traffic lights. We're anchored to shake someone's hand when he holds it out. We begin to develop anchors in childhood as we learn the cues for getting up in the morning, sitting down for a meal, responding to the moods of our parents, and observing the rituals of our families and communities.

But anchors are more than just learning the rules of behavior. Our perceptions and beliefs are anchored, too. For example, we're anchored to fear great white sharks even though few of us have ever seen one in real life, and fewer still have had an encounter with one. When we see reality-TV stars eat bugs or put their hands into a box of wriggling snakes, we're conditioned (anchored) to feel revulsion or terror even if we've never performed either activity and the bugs taste good or the snakes feel warm and soft to the touch. But for most of us our upbringing has anchored us to feel certain way.

Anchors, like Pavlov's conditioned stimuli, trigger specific states of mind, emotions, and changes to your physical state. Perhaps a particular song (auditory anchor) causes you to remember a former lover, or a particular brand of soap (olfactory anchor) reminds you of your childhood. Anchors can also be kinesthetic. Ask someone to come with you in a certain direction, put your hand behind his or her elbow, and apply gentle pressure in the direction you want him or her to go. More often than not, that gentle pressure will prompt the person to move with little or no resistance. Why? Because when we were children, our parents and teachers used that touch and gesture in the same way.

We've seen how we have both positive and negative anchors, so let's explore how to establish new anchors while disabling those we don't want. But first let's review the process, and then we'll go through an exercise that will teach you how to do it.

To embed a positive anchor, you must identify the goal you wish to achieve and then choose a behavior trigger to associate with the goal. Make the trigger something unique so you don't fire it at inappropriate times or too often, thereby weakening its emotional impact. Examples of practical triggers include snapping your fingers three times or wiggling your toes a certain way. You can engage in either of these actions without drawing too much attention to yourself, yet they're unique enough that you probably won't fire them accidentally.

Also keep in mind that you can create different anchors for different mindsets. For example, you can associate wiggling the toes on your left foot with an emotional state of relaxation, whereas snapping the fingers on your right hand might prime you to feel a sense of confidence.

Exercise

Establish Your Anchor

What does it take to create an effective anchor, and how can you develop the state of mind you want to anchor? First be aware that you don't want to anchor a mindset at the wrong time or associate a contradictory movement with a state of mind, so the physical part of the anchor should fit the specific state of mind, such as snapping your fingers or arching your back for confidence, or putting your hand gently on your forehead when you want to think more clearly.

My favorite anchor is tapping my left thumb and middle finger together three times in quick succession, which I can do almost anywhere without drawing attention to myself. I use this trigger to anchor a state of über-confidence (I chose a left-hand activation because I'm left-handed. If you're right-handed, you can choose your left hand precisely because you use it less often. Whatever makes most sense to you.) Another anchor might include tapping your knee or a particular knuckle a certain way.

Once you've decided on a unique and memorable trigger, you need to work on the mental state to which you want to anchor it. The intensity of the mindset will determine how effective your anchor will be. To create an intense state, you need to build as vivid a mental representation as possible of how you'd like to feel by using the association techniques we used earlier. It should take about twenty minutes to develop your first fully immersed state of mind. It takes time, but it's worth it. Also, the more you practice this technique, the easier it becomes.

First, picture how you will look in your chosen state. Begin with an outline of the image and introduce increasing levels of detail. Add color, and animate the image. Now step into your movie, and become part of it. Change from watching it (disassociated) to being inside it (associated).

Once you have a bright, colorful, vivid depiction, add more detail. Add layers of sound, starting with background noises, laughter, or voices and end with the sound of your own breathing. For each sense you introduce, always move from the outside in. Alternate between how the video looks and how it sounds. Experiment with switching from an associated state (in it) to a disassociated state (watching it) and back again.

Bring in scents. What can you smell? Grass? Rain? Roses? Warm apple pie à la mode? Add the aroma to the mental representation of the state you are building. Finally, what can you physically feel in your preferred state? Are you in front of a hot fire or under a cool waterfall? Are you touching soft fabric or solid stone? Can you sense drizzle falling, warm sunlight, or a faint breeze? Add whatever works best for you.

The result: a fully multi-sensory state of mind. Play your multi-sensory movie first from a disassociated perspective. Sit back, and watch it. Then play it a second time, diving into yourself and becoming immersed in your chosen state of mind. When fully immersed, perform the behavioral trigger you created.

After you're done, relax and clear your mind. Then repeat the entire process once more. The more you build your multi-sensory state of mind, the easier and quicker it becomes. Each time you immerse yourself in your desired state, activate your anchor.

After repeating this process several times, you should be able to enter the mindset just by performing your behavioral trigger. Try anchoring your preferred states of mind repeatedly over the next few weeks, and don't worry that it might fade between practice sessions. It won't.

After completing the exercise, you'll have an effective way of managing how you feel from one moment to another. If you sense you're not feeling as you want to feel—for example, just before you go into a job interview, cross the club floor to approach a potential date, reach for a cigarette, or consume that piece of cake or bottle of beer—summon the anchor. The results will amaze you. An effectively anchored mental state is stronger and faster acting than most store-bought chemicals.

Once you've successfully established an anchor, move on to the next one. Choose a different mindset-goal combination and go through the process again.

Mental Representations

Although we recall visual imagery most frequently, mental imagery involves any of the senses. Psychologist Stephen Kosslyn proposes that we use mental images to solve certain types of problems. We would visualize the objects in question and also mentally represent the images to solve the puzzle or difficulty.

In the association exercise, mental representations work because they convince our brains that the mood being felt is real. The brain doesn't distinguish between the extremely detailed daydream and the reality because it processes the information in the same way. That's why we can react to something perceived (a movie, for instance) rather than to what we actually are seeing (patterns of light moving on a screen). The brain believes it to be true.

Here's a demonstration that will prove the power of your own mind. This is easier to do with another person present, but you can do it alone, too.

Stand in the middle of a room, and extend your arms outward so that you resemble the letter T. Now rotate your right arm horizontally to the right as far as you can, and note how far that is. You can do this by aligning the fingertips of your right hand with some stationary aspect in the room, a piece of furniture or a spot on the wall, for instance. That's the measure of your ability to rotate.

Stay standing where you are, drop your arms, close your eyes, and build a mental representation of repeating the exercise. Repeat the process as described above, but spend a few seconds concentrating on each sense as you do. Imagine that you can rotate your right arm all the way around so that the back of your right hand touches the back of your left hand—farther in the representation than in reality. Once you have a clear representation in your mind, hold it for a few seconds, then relax, and open your eyes.

Close your eyes again, raise your arms into the T position again, and rotate to the right as far as you can as before. Now open your eyes, and you'll find that you've rotated farther because you imagined that you could. That's the power of mental representation.

———

Not all anchors are positive, however. As you experiment with attaching positive anchors to your success goals, you'll soon realize that you want to rid yourself of negative anchors. For example, every time you come home from work in the evening, you see the liquor cabinet, which prompts you to pour yourself a glass of wine or a slug of scotch.

The best way to rid yourself of a negative anchor—collapsing it, as the concept is known—is to replace it with an alternate and more beneficial one. The process for creating an effective replacement anchor involves repeating the technique for creating new positive anchors. Over time, the new anchor will reprogram your anticipation of the old behavior. For example, replace pouring the drink when you first come home with eating a couple of strong mints or chewing gum. That will give your brain a different trigger that you then can associate with a different mindset, such as confidence or relaxation. As a bonus in this case, the minty taste will make the alcohol less palatable, thus collapsing the negative anchor and reinforcing the positive one.

When it comes to dealing with addictions, which involve both emotional and physical cravings, you'll need to create a slightly different anchor. If you want to quit smoking, for example, you need to develop a mental representation of how it feels not to want chemical assistance—confident, proud, or relaxed for instance. When the craving hits, fire the anchor to allay it. It's similar to creating an anchor for wanting something (courage, confidence, joy, etc.), but you're developing a representation of not wanting things, which in this case are emotional and physical cravings. Whether you want more or less of something, the anchoring process works the same way.

REMEMBER . . .

* Our brains don't distinguish between detailed representations and reality.
* To anchor a mindset that you can access easily, construct a multi-sensory mental representation and associate a behavioral trigger with it.
* Develop positive anchors as shortcuts to all the beneficial mindsets you want to use.
* Note any negative anchors you have, and replace them with positive alternatives.

50

Never Give Up, Prepare to Refine

EARLIER, we talked about continuously adjusting your approach to success until you get it right. Let's take a look at some tools you can use in that regard to optimize your success.

Some people procrastinate. They find infinite excuses for not doing something now and instead taking action until later. If you're one of them, you need a means by which you can force yourself to take step one. For example, a common delaying tactic is calendar related. A lot of people decide impulsively that today just isn't the right day. Instead, they choose to address the situation sometime in the future. Our brains are comfortable accepting nearly any point in time for a target; they're not fussy. If you struggle to take that first step sooner than later, use a calendar. But don't just mark a date for when you'll take that first step. Make it more intense. Write or create a countdown on your calendar so you can see how many days remain until you take that all important step.

You'll discover that as the day of step one approaches, your brain will accept it as inevitable and begin thinking in terms of performing the action. If you keep putting off that starting date, reinforce your intentions. Never give up on your success goals, but do examine why you keep delaying. Identify what lies behind your lack of action. Once you recognize what's holding you back, refine your approach and develop a first step that your brain finds more acceptable.

What holds a lot of people back when it comes to attaining success is their resistance to taking that all-important first step, but sometimes it's a case of consistency. Frequently, it's not giving up cigarettes or alcohol so much as sticking to the plan. It's more about rejecting the first drink or deciding not to smoke the next cigarette or not eating that next cookie or just getting to the gym rather than adhering to a routine when there.

Never lose sight that you'll be successful. Don't let procrastination or a momentary slip dilute the benefits you'll experience as a result of being more successful.

One of the best ways for knowing when to refine your approach is to recognize when something isn't working. For example, if you've committed to losing a few pounds but haven't succeeded, first you need to address whether your commitment has materialized into action. Saying you're going to start your diet next week isn't enough. The further you push step one into the future, the less likely you'll have to face the action. Choose a date, commit to it, then work toward that day. When it comes, your brain will urge you to take the wanted action rather than procrastinating yet again.

Of course, if a course of action isn't working, it's important to discover why not. If a strategy isn't advancing you along your road to success, you can continue making the same effort repeatedly or stop and develop a new strategy before trying again. You also can analyze the original strategy to learn why it isn't working and refine the process rather than throwing it out altogether and starting over. If you've decided to quit smoking but every time you try you succumb each evening after work, you've succeeded in cutting out your daytime cigarettes, so you only need to refine your approach to avoid smoking in the evening.

That's a very different scenario from not being able to give up smoking at all. But how do you adjust your approach? You might avoid going to bars for a month, or you might still go but avoid drinking, or you might go to the bar for ten minutes, have one drink, and leave before lighting up. If that strategy works, you might extend your bar time to fifteen minutes, thirty minutes, or an hour. The point is that, more often than not, your whole process isn't wrong. It only needs a little tweaking. Determine which parts are working and which aren't, and re-address the latter rather than ripping up the entire plan and forcing yourself to start over from scratch.

REMEMBER . . .

* Force yourself to take the first step.
* Don't let procrastination dilute your determination to succeed.
* If a strategy isn't working, don't be afraid to look at it carefully, identify the part that isn't working, and tweak the program.

{51}

Finish to Start

D O you ever wonder what a hundred-meter sprinter thinks about as he or she settles in the blocks? In position, hands on the starting line, what goes through his or her mind? Many athletes focus on the finish line. If you remember Linford Christie winning Olympic gold in 1988, you may recall that he never blinked during the entire race. He truly focused on the goal and couldn't be distracted.

But except for short sprints, focusing on the goal is never as effective as developing a full strategy. After all, you want your brain to know where to take you. Having a clear idea of the overall goal is a great way to start, sure. But often it feels too daunting to go from where you are now to where you want to be in the future. Give your brain too much to handle, and it will go into delay mode, putting off the task because it's too large to accomplish. A more effective strategy involves a little more effort but is worth it in the end.

By now, you've created your roadmap to each aspect of your success, and it consists of what you want to achieve plus the milestones along the way to let your brain know you're on the right track. Here's a not-too-daunting process for helping you to take that critical first step.

Because your brain recognizes that small steps result in a great leap, give them to it. Identify your overall goal, then work backward to calculate how to achieve the various steps you've identified as milestones. Do it slowly and methodically, and make achieving each landmark as emotional and beneficial as reaching the overall goal. Start with step one as the first marker, plot out the rest, and you'll find that your brain won't resist setting off on your journey.

The aim of the exercise is to convince your brain that the emotional gains from taking that first step are too strong to ignore. By attaching the emotion of the overall success to each of the smaller steps along the way, you'll make each mini-goal more compelling. That increases your brain's determination to go for it.

Similarly, imagining too small a success for the overall goal will restrict how much your brain attains for you. If you want to live in a nice house, don't accept weak images from your imagination. *Nice* isn't good enough, nor will it inspire you to action. You need to include the size (big, sprawling, gigantic), preferably in square footage, along with the number of bedrooms and bathrooms and other amenities.

Whatever you want, make sure to specify your goals to your brain. Make the end point as spectacular as you can. If you aim low, you won't get anywhere. Aiming too high—along a structured route—will provide you with the maximum success your brain is able to attain.

REMEMBER . . .

* Identify your end goal and then work backward to determine intermediate milestones.
* Make that first step emotionally irresistible.
* Focus on the landmarks along the way to give your brain the success it craves.
* If you aim low, your brain won't bother making an effort, so think big.

Believe and Succeed

BY now you should understand success and how to achieve it. Motivational speaker Jim Rohn says it best: "If you really want something, you'll find a way. If you don't, then you'll find an excuse." The key to getting your brain to work toward your goals is the absolute belief that you will achieve them.

Find somewhere you can be alone for ten minutes without interruption. Close your eyes, and contemplate a specific aspect of a goal you crave. As you explore your mental representation, imagine the success of achieving the goal as if it were real. If any doubt appears or, worse still, you can't visualize what you want because you won't allow yourself to believe you can attain it, you need to conduct more rehearsals of that representation. As you do, be on the lookout for the emotional reasons that your brain isn't 100 percent committed to your goal, and work on addressing and neutralizing those.

A lack of faith in yourself often stems from an incorrectly programmed belief system, which we explored earlier. If you're having trouble getting your brain to commit fully to a particular aspect of success, write down all the reasons you must have that success, and generate a separate list of all the reasons for any negative connotations to your goal. Within this list of negatives lies the root cause of your self-doubt.

Replace each of the negatives with a positive alternative. For example, if you're stuck on believing *I'm nowhere near achieving this goal, so I'll never reach it*, you need to do some serious reframing. Your brain really can give you all the success you want; you just have to work with it rather than against it.

Also, realize that mastering one area of your life helps you master others. Having the confidence to converse with members of the opposite sex can help you speak more readily to everyone. That includes colleagues, family members, neighbors, and friends. You can trust your brain to put better words into your mouth during every conversation you have. Realigning your one fear actually amplifies the achievement in other areas.

As we reach the end of our journey, remember that, above all else, the amount of success you can achieve is directly proportional to how much you believe you can achieve. If you're convinced emotionally that you'll succeed for all the right reasons, you have a powerful tool your brain can use to help make it happen.

If you're not totally committed to achieving success, your dedication may be little more than a thought or strong desire. But desire isn't belief, so you'll need to do some fine-tuning. You may have scores of thoughts going through your mind at any given time, but none of them has real power unless it's a belief. A belief is a thought that you perceive as real.

How do you convert a meaningful thought into a motivating belief? Here's an example. Think about an elephant sitting in your refrigerator. You know that an elephant is too large to fit in the vegetable crisper, so your brain rejects the notion. But carry the image further and engage your brain in this battle. Let's assume the elephant is a small, well-trained circus animal and the refrigerator is a refrigerated trailer in front of your home. Two men pull a metal ramp from beneath the trailer, set it up, and walk the elephant up the ramp. They store the ramp, close the trailer doors, and leave. Now you've got an elephant in your refrigerator and you've turned something fanciful and unbelievable into something sensible and believable.

If you ranked your commitment to any of your success goals at less than 100 percent, identify what's causing your less than perfect score. Perhaps your success goal is unrealistic—you might want to look thirty years younger. Maybe you lack the motivation to go after a new job because you haven't given yourself a strong enough reason to set out on that journey. Whatever the reason, you need to disable the limiting influence. In looking younger, for example, consider ten years instead of thirty. It's more reasonable and, based on modern science and medicine, more achievable. If you lack the motivation to go after that new job, conjure the powerful emotional reasons for seeking alternative employment. Will the job provide an improved lifestyle, increased security, added status, or something else? What are the emotional reasons behind your answers?

Once you've considered your goal from every angle and confirmed that no hidden doubts are holding you back, you should be able to commit to it 100 percent. If so, you've just converted a thought into a belief.

As a final lever for turning thoughts into beliefs, you need to replace three small words in your vocabulary. Whenever you catch yourself using any of these words in speech, writing, or thoughts, replace them. The words are *should*, *would*, and *could*, and their replacement is *will*. "Should try harder" becomes "will try harder." "Could quit smoking" becomes "will quit smoking." And "would buy a larger house" becomes "will buy a larger house." Similarly, replace the negative forms of these—*shouldn't*, *wouldn't*, and *couldn't*—with *won't*. "I shouldn't smoke" becomes "I won't smoke," and so on. Again, you're generating a belief-oriented statement rather than a weaker thought.

Remember, the more often you perceive your goals as beliefs, the greater the likelihood that you'll make them a reality. Behind all the best efforts lies an honesty that you must reveal to yourself. Do you believe you will succeed? If so, great. If not, find out why, and change those weak, ineffective, thoughts into beliefs.

———

Now you have all the tools you need for achieving success. Like Dorothy and her ruby slippers, you had them all along—you just needed to find out for yourself how to use them. Now that you know, they'll help you make positive, meaningful, lasting changes to your health, wealth, and happiness.

REMEMBER . . .

* Anything less than a belief is a thought or desire, and neither will push you to reach your goals.
* Replace *could, should,* and *would,* with *will,* and *couldn't, shouldn't,* and *wouldn't* with *won't.*
* If you want something, you'll find a way; otherwise you'll find an excuse.
* Mastering one area of your life helps you master others.

Afterword

SO there you have it, the results of six hundred million years of evolution condensed into a single, eye-opening book. It's only through the latest advances in our understanding both of evolution itself and the human mind that this book was even possible. Writing it allowed me to adapt what I've learned in my career as a commercial psychologist into an owner's manual for the twenty-first-century human being.

I've shared a great deal of scientific facts and understanding in what I hope is plain-talking, jargon-free English. Never before have the combined findings of so many of the world's experts on the human brain come together solely for the benefit of helping you get the most from your life. Armed with what you've learned, you can now transform any aspect of your life into whatever you wish to make it. From being able to better understand and interpret the behavior of your bosses and co-workers and recognize why others behave as they do to more effectively reaching your own goals and positively influencing other's thoughts, feelings, and behavior—it's all right here at your fingertips.

Now you face a simple choice. Either slide the book onto the shelf among the other self-help books you've purchased over the years, or take what you've just read and act. This book has given you all you need to initiate massive change in your life.

Some time ago, as I sought ways of using my own brain more effectively, I came across the work of an insightful motivational speaker. After reading several of Tony Robbins's books and attending one of his four-day seminars, I made that elusive link between understanding how and why I function as I do and what to do to change my life. That's what I call the *So-What Factor*.

I recognized that the hierarchy of my own fitness indicators was driving me. I discovered that I had an almost pathological fear of debt, which was holding me back. I decided to take decisive action and control of my brain. I tackled my fear head-on by buying the biggest house that the bank would let me. The property had multiple bedrooms, magnificent views in every direction, and an eight-square-mile ancient woodland as part of the back-yard. My brand new twenty-year debt was nothing less than terrifying. But that huge debt looming over my head gave me the leverage I needed to take action. I created and implemented my own mental strategy based on what drove me emotionally.

Not long after, I learned that my company, Shopping Behaviour Xplained Ltd., had been costing more than ever to run. Initially alarmed, I soon realized that although we were spending more, we were making more. By taking on more debt, the company was reducing its initial debts more quickly. If anyone had told me beforehand to invest in and grow my company at a time when I owed more than I had ever owed in my life, I would have thought him a fool.

But as my business successes multiplied, so did those in my personal life. Like most people, I'm not a very confident public speaker, but I found myself before increasingly larger groups of people who wanted to hear my insights. I made presentations to entire departments and then advanced to public and trade conferences and even appearances on radio and TV. Once I began speaking, my brain took over. Seeing my performances played back in my mind's eye strengthened my resolve.

As a result, my business was booming less than three years later, and I'd paid off my home mortgage in full. My brain had helped me remove the debt curse that had plagued me for years. Your brains are equally remarkable in every way, so train them and use them to turn your dreams into reality.

Thanks for reading.

Acknowledgments

A S I began to think about all the people whom I should thank for their input and support to make this book possible, the list just grew and grew. So first I thank my wife, Kay, and daughter, Amy, for continuing to inspire me to work on this project day and night, month after month.

Next I thank my auntie Sylvia Jones, who many years ago introduced me to the subject of the human brain, for which I remain deeply indebted.

I must also thank D. J. Herda for helping to make this book what it is and Faye Swetky and The Swetky Agency for believing in it.

I also thank the great and the good in academia for their continual advancements in understanding how human beings operate, in particular: Robert Plutchik, Steven Pinker, and Antonio Damasio. I am grateful that their work is so accessible. I should also pay thanks to Professor James Intriligator, who is truly inspirational when it comes to interpreting our understanding of the human brain and mind.

Finally, to all those science teachers who make the fascinating subject of the human mind so boring and irrelevant to teenagers: Change. There's nothing more exciting and inspiring than understanding how we as a species developed, how we function, why we are what we are, and what we have the potential to become. Change the way you teach, spice it up.

Notes

CHAPTER 1

"... 20 million-billion calculations per second": *Scientific American: Special Report on Robots,* 2008.

73 percent water: Mitchell et al., "The Chemical Composition of the Adult Human Body and its Bearing on the Biochemistry of Growth," 625–637.

one thousand other neurons: Drachman, "Do We Have Brain to Spare?" 12.

CHAPTER 2

evolution is a process: *Oxford English Dictionary,* 2010.

our ancestors first began to speak: Uomini and Meyer, "Shared Brain Lateralization Patterns in Language and Acheulean Stone Tool Production," 2013.

CHAPTER 3

Compared to unconscious processing: Huageland, *Having Thought,* 159–60.

spent $61 billion: Marketdata Enterprises Inc., "Weight Loss Websites: A Marketing & Competitor Analysis," 2011.

CHAPTER 4

the insula . . . integrates mind and body: Martin Paulus, interview in *The New York Times,* 2007.

the insula . . . plays a starring role: Damasio, *Descartes' Error.*

CHAPTER 5

finely tuned attraction-seeking machine: Fisher, *Why We Love.*

CHAPTER 6

monkey's brains were "mirroring": Rizzolatti, *Mirrors in the Brain.*

thinking positively about getting a high grade: Phan and Taylor, "From Thought to Action: Effects of Process-Versus Outcome-Based Mental Simulations on Performance," 250–60.

CHAPTER 7

"universal set of learning principles": Clark et al., *Efficiency in Learning.*

CHAPTER 8

"Every time I learn something new . . .": Greg Daniels, *Secrets of a Successful Marriage, The Simpsons,* Series 5, Episode 22, Fox Network, 1994.

no longer than eighteen seconds: Greenfield, *The Private Life of the Brain.*

odors fade only by 3 percent: Engen, *Odor Sensation and Memory.*

CHAPTER 9:

facial image in a photograph: Ramon et al., "The Speed of Recognition of Personally Familiar Faces," 437.

soothing and exciting music: Misra, "Beneficial Effects of Musical Patterns on the Human Physiology."

cost of an item influences: Plassmann et al., "Marketing Actions Can Modulate Neural Representations of Experienced Pleasantness," 1050–54.

emotional images have more impact: Calvo and Lang, "Parafoveal Semantic Processing of Emotional Visual Scenes," 502.

listening to language: Kalat, *Biological Psychology.*

object's texture: Lindstrom, *Buy-ology.*

heavier clipboards: Ackerman et al., "Incidental Haptic Sensations Influence Social Judgments and Decisions," 1712–15.

sitting on hard surfaces: Williams and Ackerman, "The Future of Retail."

flavor . . . combination of taste and smell: Kalat, *Biological Psychology*.

women detect odors: Brizendine, *The Female Brain*.

CHAPTER 10

rapid eye movement: www.webmd.com, "What Are REM and Non-REM Sleep?"

sleep, helps our brains: Payne et al., "Memory for Semantically Related and Unrelated Declarative Information: The Benefit of Sleep, the Cost of Wake."

unconscious minds might be hard at work: de Vries et al., "The Unconscious Thought Effect in Clinical Decision Making: An Example in Diagnosis," 578–81.

Marion Jones dreamt: Barrett, *The Committee of Sleep*.

CHAPTER 11

Omega-6s help the body: Harris, "Omega-6 and Omega-3 Fatty Acids: Partners in Prevention," 125–9.

The brain needs vitamin E/ harmful free radicals: Jensen, "Vitamin E Keeps Your Brain Razor-Sharp."

found in a banana: Michael Green of Aston University, interview for *Livescience.com*, 2009.

a brisk walk: Buzan, *Use Your Perfect Memory*.

running leads to: Fred Gage of Salk Institute, *Exercise and Interaction for the Brain?*, interview for *The Naked Scientists*, 2013.

think more creatively: Colzato et al., "The Impact of Physical Exercise on Convergent and Divergent Thinking."

stroke victim's brain: Doidge, *The Brain That Changes Itself*.

mice could form new neural connections: Bochner et al., "Blocking PirB

**Up-Regulates Spines and Functional Synapses to Unlock Visual Cortical Plasticity and Facilitate Recovery from Amblyopia," 258.

human neurons making new connections: Pasko. "Neurogenesis in Adult Primate Neocortex: An Evaluation of the Evidence," 65–71.

learning producees changes: Brown University. *Study Describes Brain Changes During Learning*. ScienceDaily, October 20, 2000.

neutral state when we watch: Werman, *Living with an Aging Brain*.

"There isn't much . . .": Buchsbaum "Frontal Lobes, Basal Ganglia, Temporal Lobes—Three Sites for Schizophrenia?" 377–8.

ward off Alzheimer's disease: Wilson et al., "Relation of Cognitive Activity to Risk of Developing Alzheimer Disease," 1911–20.

CHAPTER 12

emotions as genetically based: Plutchik, *Emotions and Life*.

"the ultimate function . . .": Pinker, *How the Mind Works*.

emotion stimulates the mind: Jie et al., "Quantitative Study of Individual Emotional States in Social Networks," 132–44.

the power ratio: Erasmus of Rotterdam, article about the tension that exists between the human emotion and human reason, sixteenth century.

CHAPTER 13

emotions lie beyond: Zaltman, *How Customers Think*.

social emotions: Frank, *Mindfield*.

Social emotions aren't as old: "Fossil Reanalysis Pushes Back Origin of *Homo sapiens*," *Scientific American*.

full behavioral modernity: Mellars, "Why Did Modern Human Populations Disperse from Africa Ca. 60,000 Years Ago? A New Model," 9381–86.

emotion is an instantaneous response: Valenza, et al., "Revealing Real-Time Emotional Responses: a Personalized Assessment Based on Heartbeat Dynamics."

CHAPTER 14

95 percent of decision making: Zaltman, *How Customers Think.*

our instincts go to pleasure: Jevons, *The Theory of Political Economy.*

CHAPTER 15

I am, therefore I think: Damasio, *Descartes' Error.*

a person doesn't get depressed: Robbins, *Unlimited Power.*

Descartes had it wrong: Damasio, *Descartes' Error.*

brain-and body functionality: Pinker, *How the Mind Works.*

"The same state of mind . . .": Darwin, *The Expression of the Emotions in Man and Animals.*

CHAPTER 16

effective emotional management: Ostroff and Judge, *Perspectives on Organizational Fit.*

"Every day, in every way, I'm getting better": Coué, *De la suggestion et de ses applications.*

CHAPTER 18

Emotions embed memories: Kensinger, "Remembering the Details: Effects of Emotion," 99–113.

Remembering the good times: Walker et al., "Life Is Pleasant—and Memory Helps to Keep It That Way!" 203.

CHAPTER 18

emotions aroused: Kensinger, "Remembering the Details: Effects of Emotion," 99–113.

memories are pleasant: Walker et al., "Life Is Pleasant—and Memory Helps to Keep It That Way!" 203.

CHAPTER 19

music can arouse feelings: Salimpoor et al., "Anatomically Distinct Dopamine Release During Anticipation and Experience of Peak Emotion to Music," 257–62.

CHAPTER 21

men "know" something/hip size/growing fetus: Lassek et al., "Eternal Curves."

CHAPTER 23

Fear is both a noun and a verb: *The Oxford English Dictionary.*

Harland Sanders took a job: Ozersky, *Colonel Sanders and the American Dream.*

Sander's restaurant business/By 1963: Jakle and Sculle, *Fast Food.*

finger-lickin' $2 million: Smith, *Fast Food and Junk Food.*

a lifetime salary: Robert Cottreli, "Obituary: Colonel Sanders," *Financial Times,* December 17, 1980.

$23 billion in revenue: Yum! Brands, 2014.

". . . go back to driving a truck": Eikleberry, *The Career Guide for Creative and Unconventional People.*

CHAPTER 24

imposter syndrome: Clance and Imes, "The Imposter Phenomenon in High Achieving Women: Dynamics and Therapeutic Intervention," 241.

CHAPTER 25

whales and wolves: Wilson, *Sociobiology.*

hardwired to band together: Tomasello et al., "Two Key Steps in the Evolution of Human Cooperation," 673–92.

babies cry at the sound: Dondi et al., "Can Newborns Discriminate Between Their Own Cry and the Cry of Another Newborn Infant?" 418–26.

behave differently in groups: Adcock, *Supermarket Shoppology.*

CHAPTER 26

Scientists have discovered: Fisher, *Why We Love.*

"Men like an exaggerated female form": Cox, *Stiletto.*

physical presentation: Matts et al., "Color homogeneity and visual perception of age, health, and attractiveness of female facial skin," 977–84.

a woman's scent: Havlíček et al., "Non-Advertized Does Not Mean Concealed: Body Odour Changes Across the Human Menstrual Cycle," 81–90.

". . . no such thing as a safe tan": U.S. Food and Drug Administration, "The Risks of Tanning," 2013. See also: Brenner and Hearing, "The Protective Role of Melanin Against UV Damage in Human Skin," 539–49.

CHAPTER 27

how quickly we arrive: Willis and Todorov, "First Impressions Making Up Your Mind after a 100-ms Exposure to a Face," 592–98.

"You've got just seven seconds . . .": Ailes, *You Are the Message.*

ideal amount of makeup: Jones et al., "Miscalibrations in Judgements of Attractiveness with Cosmetics," ahead-of-print.

CHAPTER 29

55 to 70 percent: Mehrabian, *Silent Messages.*

CHAPTER 30

"The young and . . ."/happiness visible from farther/Humans indicate dislike: Darwin, *The Expression of the Emotions in Man and Animals.*

branchial arches: Darwin, *The Voyage of the Beagle.*

survival technique: Sagan, *Cosmos.*

facial expressions reveal more: Ekman and Rosenberg, *What the Face Reveals.*

"The expression a woman wears": Carnegie, *How to Win Friends and Influence People.* 2006

women first looked at images: Jones et al., "Miscalibrations in Judgements of Attractiveness with Cosmetics," ahead-of-print.

angle of your face: Burke and Sulikowski, "A New Viewpoint on the Evolution of Sexually Dimorphic Human Faces," 573–85.

1,084 heterosexual men: Tracy and Bealle, "Happy Guys Finish Last: The Impact of Emotion Expressions on Sexual Attraction," 1379.

men consistently find women in makeup: Guéguen, "The Effect of Facial Makeup on the Frequency of Drivers Stopping for Hitchhikers 1," 97–101.

detecting what type of smile: Shopping Behaviour Xplained Ltd, real smile vs. polite smile research, 2012.

CHAPTER 31

OCEAN: Triandis et al., "Cultural Influences on Personality," 133–60.

after the age of thirty: Kelly, "Consistency of the Adult Personality," 659–81.

"the desire for self-fulfillment . . .": Maslow, "A Theory of Human Motivation," 370–96.

"There is very little . . .": James, "The Importance of Individuals."

general intelligence correlates/ root of consumer capitalism: Miller, *Must-Have*.

lack of openness/ Conscientiousness goes hand in hand: Miller, *Spent*.

train wasn't needed: Wilson, *The Social Conquest of Earth*.

admirable traits in a male/ man is evaluating: Kenrick and Griskevicius, *The Rational Animal*.

average scores on the text: Bouchard and McGue, Matt, "Genetic and Environmental Influences on Human Psychological Differences," 4–45.

CHAPTER 32
women like to share their problems: Pease and Pease, *Why Men Don't Listen and Women Can't Read Maps*.

CHAPTER 33
long-term benefits of dual coding: Paivio and Foth, "Imaginal and Verbal Mediators and Noun Concreteness in Paired-Associate Learning: The Elusive Interaction," 384–90.

CHAPTER 35
neuro-linguistic programming: Bandler and Grinder, *Frogs into Princes*.

"NLP represents . . .": Witkowski, "Thirty-Five Years of Research on Neuro-Linguistic Programming. NLP Research Data Base. State of the Art or Pseudoscientific Decoration?" 58–66.

"Mirroring as an important . . .": Kavanagh et al., "When It's an Error to Mirror: The Surprising Reputational Costs of Mimicry," 1274–76.

food servers mirrored: Gladwell, *Blink*.

Representational systems are the inner: Charalambos, "Involving the Concepts of the Individual Learning Style and the Self-Esteem in Redefining Quality in Adult Education."

preferred representational system: Pantin, "The Relationship Between Subjects' Predominant Sensory Predicate Use, Their Preferred Representational System and Self-Reported Attitudes Towards Similar Versus Different Therapist-Patient Dyads."

people use all of their senses: Sharpley, "Predicate Matching in NLP: A Review of Research on the Preferred Representational System," 238.

CHAPTER 36
address these types of phobias: Morris et al., "BF Skinner's Contributions to Applied Behavior Analysis," 99.

CHAPTER 38
form in a number of ways/"Common sense . . .": Halligan and Aylward, *The Power of Belief*.

"Procrastination is . . .": Kiam, *Going for It!*

CHAPTER 41
limit our success: Weger and Loughnan, "Mobilizing Unused Resources: Using the Placebo Concept to Enhance Cognitive Performance," 23–8.

good-news bias: Sharot, *The Optimism Bias*.

left and right sides of the brain: Sperry, *Science and Moral Priority*.

CHAPTER 42
"a more or less . . .": Andrews, "Habit," 121–49.

CHAPTER 46
10 percent: Sharps et al., "It's the End of the World and They Don't Feel Fine."

CHAPTER 47
"repeating the same . . .": World Service Office, *Narcotics Anonymous*.

"I made 5,127...": Benett and Maynard, *The Talent Mandate.*
"Try again. Fail again. Fail Better.": Beckett, *Worstward Ho.*

CHAPTER 48

scanned the brains/"This is scientific...": Kosslyn et al., "Hypnotic Visual Illusion Alters Color Processing in the Brain," 1279–84.

hypnotism gives the right brain: Jaynes, *The Origin of Consciousness in the Breakdown of the Bicameral Mind.*

CHAPTER 49

use mental images: Kosslyn, *Clear and to the Point.*

CHAPTER 52

"If you really want something...": Rohn, *7 Strategies for Wealth & Happiness.*

Select Bibliography

BOOKS

Adcock, Phillip. *Supermarket Shoppology.* Staffordshire: Shopping Behaviour Xplained Ltd, 2011.

Ailes, Roger. *You Are the Message.* New York: Doubleday, 1988.

Bandler, Richard, and John Grinder. *Frogs into Princes: Neuro Linguistic Programming.* Eden Grove Editions, 1990.

Barrett, Deirdre. *The Committee of Sleep: How Artists, Scientists, and Athletes Use Their Dreams for Creative Problem Solving—And How You Can Too.* New York: Crown Publishing, 2001.

Beckett, Samuel. *Worstward Ho.* London: Faber & Faber, 1983.

Benett, Andrew, and Barksdale Maynard, *The Talent Mandate: Why Smart Companies Put People First.* New York: Palgrave Macmillan, St. Martin's Press, 2013.

Brizendine, Louann. *The Female Brain.* New York: Broadway Books, Random House 2006.

Buzan, Tony. *Use Your Perfect Memory: Dramatic New Techniques for Improving Your Memory.* New York: Penguin Group, 1991.

Carnegie, Dale. *How to Win Friends and Influence People.* New York: Simon & Schuster, 2006.

Clark, Ruth, Frank Nguyen, and John Sweller. *Efficiency in Learning: Evidence-Based Guidelines to Manage Cognitive Load.* San Francisco: Pfeiffer, John Wiley & Sons, 2006.

Coué, Émile. *De la suggestion et de ses applications a la thérapeutique.* Paris: Octave Doin, Éditeur, 1888.

Cox, Caroline. *Stiletto.* New York: Collins Design, 2004.

Damasio, Antonio. *Descartes' Error: Emotion, Reason, and the Human Brain.* New York: Penguin Group, 1994.

Darwin, Charles. *The Expression of the Emotions in Man and Animals.* New York: D. Appleton and Company, 1886.

———. *The Voyage of the Beagle: Charles Darwin's Journal of Researches.* New York: Penguin Classics, 1989.

Doidge, Norman. *The Brain That Changes Itself: Stories of Personal Triumph from the Frontiers of Brain Science.* New York: Penguin Group, 2007.

Ekman, Paul, and Erika L. Rosenberg. *What the Face Reveals: Basic and Applied Studies of Spontaneous Expression Using the Facial Action Coding System (FACS).* New York: Oxford University Press, 2005.

Eikleberry, Carol. *The Career Guide for Creative and Unconventional People.* Berkeley: Ten Speed Press, 2007.

Engen, Trygg. *Odor Sensation and Memory.* New York: Praeger Publishers, Greenwood Publishing, 1991.

Fisher, Helen. *Why We Love: The Nature and Chemistry of Romantic Love.* New York: Owl Books, Henry Holt and Company, 2004.

Frank, Lone. *Mindfield: How Brain Science Is Changing Our World.* Oxford: Oneworld Publications, 2009.

Gladwell, Malcolm. *Blink: The Power of Thinking Without Thinking.* New York: Little Brown, 2005.

Greenfield, Susan. *The Private Life of the Brain: Emotions, Consciousness, and the Secret of the Self.* New York: John Wiley & Sons, 2000.

Halligan, Peter, and Mansel Aylward. *The Power of Belief: Psychological Influence on Illness, Disability, and Medicine*. Oxford: Oxford University Press, 2006.

Huageland, John. *Having Thought: Essays in the Metaphysics of Mind*. Cambridge: Harvard University Press, 1998, 159–60.

Jakle, John, and Keith Sculle, *Fast Food: Roadside Restaurants in the Automobile Age*. Baltimore: John Hopkins University Press, 1999.

Jaynes, Julian. *The Origin of Consciousness in the Breakdown of the Bicameral Mind*. New York: Houghton Mifflin, 2000.

Jevons, William Stanley. *The Theory of Political Economy*. London: Macmillan, 1871.

Kalat, James. *Biological Psychology*. Belmont: Wadsworth, 2009.

Kenrick, Douglas T., and Vladas Griskevicius. *The Rational Animal: How Evolution Made Us Smarter Than We Think*. New York: Basic Books, 2013.

Kiam, Victor. *Going for It!: How to Succeed As an Entrepreneur*. New York: William Morrow and Company, HarperCollins Publishers, 1986.

Kosslyn, Stephen M. *Clear and to the Point: 8 Psychological Principles for Compelling PowerPoint Presentations*. Oxford: Oxford University Press, 2007.

Lindstrom, Martin. *Buyology: Truth and Lies About Why We Buy*. New York: Doubleday, Random House, 2008.

Mehrabian, Albert. *Silent Messages: Implicit Communication of Emotions and Attitudes*. Boston: Wadsworth Publishing, 1972.

Miller, Geoffrey. *Must-Have: The Hidden Instincts Behind Everything We Buy*. New York: Vintage, Penguin Group, 2010.

———. *Spent: Sex, Evolution and the Secrets of Consumerism*. New York: Viking, Penguin Group, 2009.

Ostroff, Cheri, and Timothy A. Judge. *Perspectives on Organizational Fit*. New York: Psychology Press, Taylor & Francis, 2007.

Oxford Dictionaries. *The Oxford English Dictionary*. Oxford: Oxford University Press, 2010.

Ozersky, Josh. *Colonel Sanders and the American Dream*. Austin: University of Texas Press, 2012.

Pease, Allan, and Barbara Pease. *Why Men Don't Listen and Women Can't Read Maps: How We're Different and What to Do About It*. New York: Broadway, 1998.

Pinker, Steven. *How the Mind Works*. New York: W. W. Norton & Company, 1997.

Plutchik, Robert. *Emotions and Life: Perspectives from Psychology, Biology, and Evolution*. Washington, DC: American Psychological Association, 2002.

Rizzolatti, Giacomo. *Mirrors in the Brain: How Our Minds Share Actions and Emotions*. Oxford: Oxford University Press, 2006.

Robbins, Tony. *Unlimited Power: The New Science of Personal Achievement*. New York: Free Press, Simon & Schuster, 1986.

Rohn, Jim. *7 Strategies for Wealth & Happiness: Power Ideas from America's Foremost Business Philosopher*. New York: Three Rivers Press, Random House, 1996.

Sagan Carl. *Cosmos: The Story of Cosmic Evolution, Science and Civilization*. Abacus, 1983.

Sharot, Tali. *The Optimism Bias: A Tour of the Irrationally Positive Brain*. New York: Vintage Books, Random House, 2012.

Smith, Andrew F. *Fast Food and Junk Food: The Dark Side of the All-American Meal*. Santa Barbara: ABC-CLIO, 2012.

Snell, Bruno. *The Discovery of the Mind*. Oxford: Basil Blackwell, 1952.

Sperry, Roger. *Science and Moral Priority: Merging Mind, Brain and Human Values*. New York: Columbia University Press, 1985.

Werman, Robert. *Living with an Aging Brain: A Self-Help Guide for Your Senior Years*. Tel Aviv: Freund Publishing, 2003.

Wilson, Edward O. *Sociobiology: The New Synthesis*. Cambridge: Belknap Press of Harvard University Press, 2000.

———. *The Social Conquest of Earth*. New York: Liveright, W. W. Norton, 2013.

World Service Conference Literature Sub-Committee of Narcotics Anonymous. *Narcotics Anonymous: The Basic Text*. World Service Conference Literature Sub-Committee of Narcotics Anonymous, 1981.

Zaltman, Gerald. *How Customers Think: Essential Insights into the Mind of the Market*. Boston: Harvard Business School Press, 2003.

REPORTS AND ARTICLES

Ackerman, Joshua M., Christopher C. Nocera, and John A. Bargh. "Incidental Haptic Sensations Influence Social Judgments and Decisions." *Science* 328, no. 5986 (2010): 1712–15.

Andrews, Benjamin Richard. "Habit." *The American Journal of Psychology* 14, no. 2 (1903): 121–49.

Barnett, Lincoln Barnett. "The Universe and Dr. Einstein: Part II." *Harper's Magazine* 196 (1948): 473.

Bochner, David N., Richard W. Sapp, Jaimie D. Adelson, Siyu Zhang, Hanmi Lee, Maja Djurisic, Josh Syken, Yang Dan, and Carla J. Shatz. "Blocking PirB Up-Regulates Spines and Functional Synapses to Unlock Visual Cortical Plasticity and Facilitate Recovery from Amblyopia." *Science Translational Medicine* 6, no. 258 (2014): 258. doi: 10.1126/scitranslmed.3010157.

Bouchard, Thomas J., and Matt McGue, "Genetic and Environmental Influences on Human Psychological Differences."

Journal of Neurobiology 54, no. 1 (2003): 4–45. doi: 10.1002/neu.10160.

Brenner, Michaela, and Vincent J. Hearing. "The Protective Role of Melanin Against UV Damage in Human Skin." *Photochemistry and Photobiology* 84, no. 3 (2008): 539–49. doi: 10.1111/j.1751-1097.2007.00226.x.

Buchsbaum, Monte S. "Frontal Lobes, Basal Ganglia, Temporal Lobes—Three Sites for Schizophrenia?" *Schizophrenia Bulletin* 16, no. 3 (1990): 377–8.

Burke, Darren, and Danielle Sulikowski. "A New Viewpoint on the Evolution of Sexually Dimorphic Human Faces." *Evolutionary Psychology: An International Journal of Evolutionary Approaches to Psychology and Behavior* 8, no. 4 (2009): 573–85.

Calvo, Manuel G., and Peter J. Lang. "Parafoveal Semantic Processing of Emotional Visual Scenes." *Journal of Experimental Psychology: Human Perception and Performance* 31, no. 3 (2005): 502.

Charalambos, Tsiros. "Involving the Concepts of the Individual Learning Style and the Self-Esteem in Redefining Quality in Adult Education." *Educating the Adult Educators*, ReNAdE&T e-book Conference Proceedings, University of Peloponnese, Greece, 2009.

Clance, Pauline R., and Suzanne A. Imes. "The Imposter Phenomenon in High Achieving Women: Dynamics and Therapeutic Intervention." *Psychotherapy: Theory, Research & Practice* 15, no. 3 (1978): 241.

Colzato, Lorenza S., Ayca Szapora, Justine N. Pannekoek, and Bernhard Hommel. "The Impact of Physical Exercise on Convergent and Divergent Thinking." *Frontiers in Human Neuroscience* 7 (2013).

de Vries, Marieke, Cilia LM Witteman, Rob W. Holland, and Ap Dijksterhuis. "The Unconscious Thought Effect in Clinical Decision Making: An Example in Diagnosis." *Medical Decision Making* 30, no. 5 (2010): 578–81.

Dondi, Marco, Francesca Simion, and Giovanna Caltran. "Can Newborns Discriminate Between Their Own Cry and the Cry of Another Newborn Infant?" *Developmental Psychology* 35, no. 2 (1999): 418–26.

Drachman, David A. "Do We Have Brain to Spare?" *Neurology* 64, no. 12, (2004–5): 12.

Guéguen, Nicolas, and Lubomir Lamy. "The Effect of Facial Makeup on the Frequency of Drivers Stopping for Hitchhikers 1." *Psychological Reports* 113, no. 1 (2013): 97-101.

Harris, William. "Omega-6 and Omega-3 Fatty Acids: Partners in Prevention." *Current Opinion in Clinical Nutrition & Metabolic Care* 13, no. 2 (2010): 125–9.

Havlíček, Jan, Radka Dvořáková, Luděk Bartoš, and Jaroslav Flegr. "Non-Advertized Does Not Mean Concealed: Body Odour Changes Across the Human Menstrual Cycle." *Ethology* 112, no. 1 (2006): 81–90.

James, William. "The Importance of Individuals." Essay, 1890.

Jensen, Aaron W., "Vitamin E Keeps Your Brain Razor-Sharp." *Life Enhancement* November 2002.

Jones, Alex L., Robin SS Kramer, and Robert Ward. "Miscalibrations in Judgements of Attractiveness with Cosmetics." *The Quarterly Journal of Experimental Psychology* ahead-of-print (2014): 1–9.

Kavanagh, Liam, Christopher Suhler, Patricia Smith Churchland, and Piotr Winkielman. "When It's an Error to

Mirror: The Surprising Reputational Costs of Mimicry." *Psychological Science* 22, no. 10 (2011): 1274–76.

Kelly, E. Lowell. "Consistency of the Adult Personality." *American Psychologist* 10, no. 11 (1955): 659–81.

Kensinger, Elizabeth A. "Remembering the Details: Effects of Emotion." *Emotion Review* 1, no. 2 (2009): 99–113.

Kosslyn, Stephen M., William L. Thompson, Maria F. Costantini-Ferrando, Nathaniel M. Alpert, and David Spiegel. "Hypnotic Visual Illusion Alters Color Processing in the Brain." *American Journal of Psychiatry* 157, no. 8 (2000): 1279–84.

Lassek, Will, Steve Gaulin, and Hara Estroff Marano. "Eternal Curves." *Psychology Today*, July 2012.

Market Data Enterprises Inc. "Weight Loss Websites: A Marketing & Competitor Analysis." MarketResearch.com, January 2011.

Maslow, Abraham. "A Theory of Human Motivation." *Psychological Review* 50, no. 4 (1943): 370–96.

Matts, Paul J., Bernhard Fink, Karl Grammer, and Maria Burquest. "Color homogeneity and visual perception of age, health, and attractiveness of female facial skin." *Journal of the American Academy of Dermatology* 57, no. 6 (2007): 977–84.

Mellars, Paul. "Why Did Modern Human Populations Disperse from Africa Ca. 60,000 Years Ago? A New Model." *Proceedings of the National Academy of Sciences* 103, no. 25 (2006): 9381–86.

Misra, Shveata. "Beneficial Effects of Musical Patterns on the Human Physiology." Paper presented at the International Seminar on "Creating & Teaching Music Patterns," Rabindra

Bharati University, Rajasthan, India, December 16–18 2013.

Mitchell, H. H., T. S. Hamilton, F. R. Steggerda, and H. W. Bean. "The Chemical Composition of the Adult Human Body and its Bearing on the Biochemistry of Growth." *Journal of Biological Chemistry* 158, (1945): 625–37.

Morris, Edward K., Nathaniel G. Smith, and Deborah E. Altus. "BF Skinner's Contributions to Applied Behavior Analysis." *The Behavior Analyst* 28, no. 2 (2005): 99.

Morris, Martha Clare, Denis A. Evans, Julia L. Bienias, Christine C. Tangney, and Robert S. Wilson. "Vitamin E and Cognitive Decline in Older Persons." *Archives of Neurology* 59, no. 7 (2002): 1125–32.

Paivio, Allan, and Dennis Foth. "Imaginal and Verbal Mediators and Noun Concreteness in Paired-Associate Learning: The Elusive Interaction." *Journal of Verbal Learning and Verbal Behavior* 9, no. 4 (1970): 384–90.

Pantin, Hilda Maria. "The Relationship Between Subjects' Predominant Sensory Predicate Use, Their Preferred Representational System and Self-Reported Attitudes Towards Similar Versus Different Therapist-Patient Dyads." (1982).

Payne, Jessica D., Matthew A. Tucker, Jeffrey M. Ellenbogen, Erin J. Wamsley, Matthew P. Walker, Daniel L. Schacter, and Robert Stickgold. "Memory for Semantically Related and Unrelated Declarative Information: The Benefit of Sleep, the Cost of Wake." *PloS one* 7, no. 3 (2012): e33079.

Phan, Lien B., and Shelley E. Taylor, "From Thought to Action: Effects of Process-Versus Outcome-Based Mental Simulations on Performance." *Personality and Social Psychology Bulletin* 25, no. 2 (1999): 250–60. doi: 10.1177/0146167299025002010.

Plassmann, Hilke, John O'Doherty, Baba Shiv, and Antonio Rangel. "Marketing Actions Can Modulate Neural Representations of Experienced Pleasantness." *Proceedings of the National Academy of Sciences* 105, no. 3 (2008): 1050–54.

Rakic, Pasko. "Neurogenesis in Adult Primate Neocortex: An Evaluation of the Evidence." *Nature Reviews Neuroscience* 3 (2002): 65–71.

Ramon, Meike, Stephanie Caharel, and Bruno Rossion. "The Speed of Recognition of Personally Familiar Faces." *Perception-London* 40, no. 4 (2011): 437.

Salimpoor, Valorie N., Mitchel Benovoy, Kevin Larcher, Alain Dagher, and Robert J. Zatorre. "Anatomically Distinct Dopamine Release During Anticipation and Experience of Peak Emotion to Music." *Nature Neuroscience* 14, no. 2 (2011): 257–62.

Scientific American. "Fossil Reanalysis Pushes Back Origin of *Homo sapiens*." *Scientific American,* February 17, 2005.

Sharpley, Christopher F. "Predicate Matching in NLP: A Review of Research on the Preferred Representational System." *Journal of Counseling Psychology* 31, no. 2 (1984): 238.

Sharps, Matthew J., Schuyler W. Liao, and Megan R. Herrera, "It's the End of the World and They Don't Feel Fine." *Skeptical Inquirer* 37 no. 1 (2013).

Tang, Jie, Yuan Zhang, Jimeng Sun, Jinhai Rao, Wenjing Yu, Yiran Chen, and Alvis Cheuk M. Fong. "Quantitative Study of Individual Emotional States in Social

Networks." *IEEE Transactions on Affective Computing* 3, no. 2 (2012): 132–44.

Tomasello, Michael, Alicia P. Melis, Claudio Tennie, Emily Wyman, and Esther Herrmann. "Two Key Steps in the Evolution of Human Cooperation." *Current Anthropology* 53, no. 6 (2012): 673–92.

Tracy, Jessica L., and Alec T. Beall. "Happy Guys Finish Last: The Impact of Emotion Expressions on Sexual Attraction." *Emotion* 11, no. 6 (2011): 1379.

Triandis, Harry C., and Eunkook M. Suh. "Cultural Influences on Personality." *Annual Review Of Psychology* 53, no. 1 (2002): 133–60.

Uomini, Natalie Thaïs, and Georg Friedrich Meyer. "Shared Brain Lateralization Patterns in Language and Acheulean Stone Tool Production: A Functional Transcranial Doppler Ultrasound Study." *PLoS one* 8(8): e72693 (2013). doi:10.1371/journal.pone.0072693.

Valenza, Gaetano, Luca Citi, Antonio Lanatá, Enzo Pasquale Scilingo, and Riccardo Barbieri. "Revealing Real-Time Emotional Responses: a Personalized Assessment Based on Heartbeat Dynamics." *Scientific Reports* 4 (2014).

Walker, W. Richard, John J. Skowronski, and Charles P. Thompson. "Life Is Pleasant—and Memory Helps to Keep It That Way!" *Review of General Psychology* 7, no. 2 (2003): 203.

Weger, Ulrich W., and Stephen Loughnan. "Mobilizing Unused Resources: Using the Placebo Concept to Enhance Cognitive Performance." *The Quarterly Journal of Experimental Psychology* 66, no. 1 (2013): 23–8.

Williams, Lawrence, and Joshua Ackerman. "The Future of Retail." *A Harvard Business Review Insight Center,* 2012.

Willis, Janine, and Alexander Todorov. "First Impressions Making Up Your Mind after a 100-ms Exposure to a Face." *Psychological Science* 17, no. 7 (2006): 592–98.

Wilson, Robert S., Paul A. Scherr, Julie A. Schneider, Yuxiao Tang, and David A. Bennett. "Relation of Cognitive Activity to Risk of Developing Alzheimer Disease." *Neurology* 69, no. 20 (2007): 1911–20.

Witkowski, Tomasz. "Thirty-Five Years of Research on Neuro-Linguistic Programming. NLP Research Data Base. State of the Art or Pseudoscientific Decoration?" *Polish Psychological Bulletin* 41, no. 2 (2010): 58–66.

WEBSITES

www.alzheimersprevention.org
www.fda.gov
www.ft.com
www.life-enhancement.com
www.livescience.com
www.neurology.org
www.nytimes.com
www.sciencedaily.com
www.scientificamerican.com
www.thefreedictionary.com
www.thenakedscientists.com
www.ualberta.ca
www.webmd.com

Index